Tourism, Ethnic Diversity and the City

It is hard to imagine urban tourism today without immigrants. Immigrants often provide the cheap labour or the enterpreneurial drive for the urban tourism industry. Moreover, their real or imagined cultural expressions are increasingly discernible amongst the 'objects' of urban tourism. More and more travellers, leisure seekers and business investors in gateway cities are indulging in ethnocultural events and festivals and are gravitating to centres of immigrant ethnic commerce. The urban tourist economy is thus becoming one of the interfaces between immigrants from all strata of society and the wider economy.

Tourism, Ethnic Diversity and the City explores the manifestations of ethnic diversity that have been commodified by immigrants in gateway cities and it asks how these expressions of culture can be transformed into vehicles for further developing the urban tourism economy. The primary focus is on the role of immigrant entrepreneurs and workers in the emerging urban tourism industry and on their interactions with other players in that industry. The relative roles of public, private and civil society actors are important points of attention. By addressing these issues from an interdisciplinary and comparative perspective, a more thorough understanding of the structural dynamics of immigrants' commercial manifestations of ethnic diversity is sought. The book further examines how such activities serve to integrate immigrants into the knowledge economy and how they impact upon urban socioeconomic development as a whole.

Tourism, Ethnic Diversity and the City clearly explores the frontiers of knowledge on the interrelationship between tourism, migration, ethnic diversity and place. It investigates new theoretical insights and challenges for empirical research using case studies drawn from several advanced economies in Europe, North America and Australia.

Jan Rath is Professor of Urban Sociology and Academic Director of the Institute for Migration and Ethnic Studies at the University of Amsterdam.

Contemporary Geographies of Leisure, Tourism and Mobility
Series Editor: C. Michael Hall
Professor at the Department of Tourism, University of Otago,
New Zealand

The aim of this series is to explore and communicate the intersections and relationships between leisure, tourism and human mobility within the social sciences.

It will incorporate both traditional and new perspectives on leisure and tourism from contemporary geography, e.g. notions of identity, representation and culture, while also providing for perspectives from cognate areas such as anthropology, cultural studies, gastronomy and food studies, marketing, policy studies and political economy, regional and urban planning, and sociology, within the development of an integrated field of leisure and tourism studies.

Also, increasingly, tourism and leisure are regraded as steps in a continuum of human mobility. Inclusion of mobility in the series offers the prospect to examine the relationship between tourism and migration, the sojourner, educational travel, and second home and retirement travel phenomena.

The series comprises two strands:

Contemporary Geographies of Leisure, Tourism and Mobility aims to address the needs of students and academics, and the titles will be published in hardback and paperback. Titles include:

The Moralisation of Tourism
Sun, sand … and saving the world?
Jim Butcher

The Ethics of Tourism Development
Mick Smith and Rosaleen Duffy

Tourism in the Caribbean
Trends, development, prospects
Edited by David Timothy Duval

Qualitative Research in Tourism
Ontologies, epistemologies and
methodologies
*Edited by Jenny Phillimore and
Lisa Goodson*

The Media and the Tourist Imagination
Converging cultures
*Edited by David Crouch, Rohna Jackson
and Felix Thompson*

**Tourism and Global Environmental
Change**
Ecological, social, economic and political
interrelationships
*Edited by Stefan Gössling and
C. Michael Hall*

Routledge Studies in Contemporary Geographies of Leisure, Tourism and Mobility is a forum for innovative new research intended for research students and academics, and the titles will be available in hardback only. Titles include:

Tourism, Ethnic Diversity and the City

Edited by
Jan Rath

Routledge
Taylor & Francis Group

NEW YORK AND LONDON

First published 2007 by Routledge
207 Madison Ave, New York, NY 10016

Simultaneously published in the UK
by Routledge
2 Park Square, Milton Park, Abingdon, Oxon OX14 4RN

Routledge is an imprint of the Taylor & Francis Group

Typeset in Times New Roman by Keyword Group Ltd, Wallington
Printed and bound in Great Britain by Biddles Ltd, King's Lynn

British Library Cataloguing in Publication Data
A catalogue record for this book is available from the British Library

Library of Congress Cataloging in Publication Data

Tourism, ethnic diversity, and the city / [edited by] Jan Rath.
p. cm. — (Routledge contemporary geographies of leisure, tourism and
mobility)
Includes bibliographical references and index.
ISBN 0–415–33390–3 (hardcover : alk. paper) 1. Tourism. 2. Tourism –
Social aspects. 3. Cities and towns. 4. Ethnic groups – Economic aspects.
5. Minority business enterprises. I. Rath, Jan, 1956- II. Title III. Series:
Contemporary geographies of leisure, tourism and mobility.

G155.A1T59175 2007
338.4'791 – dc22
2006014482

ISBN10: 0–415–33390–3 (hbk)
ISBN10: 0–203–41386–5 (ebk)

ISBN 13: 978–0–415–33390–0 (hbk)
ISBN 13: 978–0–203–41386–9 (ebk)

To Mieke van de Rhee

Contents

Illustrations

Figures

Tables

Contributors

Gastón Alonso received his PhD from the University of California, Berkeley, and is Assistant Professor in the Department of Political Science at Brooklyn College, City University of New York. He is a specialist in urban politics, migration studies and US politics. His book *In the Shadow of the State: Locating migration in transnational Miami* is forthcoming. It is based on five years of fieldwork conducted in Miami–Dade County, Florida, and analyses the political and economic incorporation of immigrants living in global cities in the United States.

Volkan Aytar is a doctoral candidate in the Department of Sociology, State University of New York, Binghamton, and a programme officer for the Turkish Economic and Social Studies Foundation (TESEV) in Istanbul. His academic interests include the forms of employment in Istanbul's entertainment and tourism establishment, the social construction of spaces of music, and consumption and its symbolic meanings for the identities of various classes and class fractions.

Hugh Bartling is Assistant Professor of Public Policy Studies at DePaul University, Chicago. His research focuses on policies of urban and suburban development in North America and Europe and on the intersections between public policy and cultural identity. He is co-editor of *Suburban Sprawl: Culture, theory and politics* (2003) and has published multiple articles on planned communities in the United States.

Jock Collins is Professor of Economics at the University of Technology (UTS) in Sydney. His research interests centre on the interdisciplinary study of immigration and cultural diversity in the economy and society. His recent research has been on Australian immigration, ethnic crime, immigrant entrepreneurship, ethnic precincts and tourism, and the social use of ethnic heritage and built environment. He has published extensively in the field, with eight books, numerous articles in national and international journals, and chapters in books. His last book, co-authored with others, is *Bin Laden in the Suburbs: Criminalising the Arabic 'Other'* (2004).

Susan S. Fainstein is Professor of Urban Planning at the Graduate School of Architecture, Planning and Preservation, Columbia University, New York. She has written extensively on urban redevelopment, urban theory and comparative public policy. Among other books, she is co-editor of *The Tourist City* (1999) and *Cities and Visitors* (2003) and author of *The City Builders: Property development in New York and London, 1980–2000* (2001).

Kevin Fox Gotham is Associate Professor of Sociology at Tulane University, New Orleans. His interests are in social theory, race and ethnicity, urban sociology and comparative historical sociology. His empirical research focuses on metropolitan development, the causes and consequences of racial residential segregation, and the historical development of US federal housing policy. He is the author of *Race, Real Estate and Uneven Development: the Kansas City experience, 1900–2000* (2002) and is editor of *Critical Perspectives on Urban Redevelopment* (2001). He is currently exploring the rise and dominance of tourism in New Orleans and the globalization of US real estate industry.

C. Michael Hall, at the time of writing, was a professor in the Department of Tourism, University of Otago, New Zealand, and a *gästforskare* at the Department of Social and Economic Geography, Umeå University, Sweden. He is co-editor of *Current Issues in Tourism* and has wide-ranging interests in tourism, mobility, regional development and environmental history. Recent publications include *Tourism: Rethinking the social science of mobility* (2005), *Tourism and Postcolonialism* (Routledge 2004), *Tourism, Mobility and Second Homes* (2004), *Companion to Tourism* (2004) and *Tourism and Migration* (2002).

Marilyn Halter is Professor of History and Director of the American and New England Studies Program at Boston University, where she is also a Research Associate at the Institute on Culture, Religion and World Affairs (CURA). An interdisciplinary scholar, she specializes in the history and sociology of American immigration, race, ethnicity, entrepreneurship and consumer culture. She is the author or editor of numerous articles, book chapters and books, including *Between Race and Ethnicity: Cape Verdean American Immigrants, 1860–1965* (1993), *New Migrants in the Marketplace: Boston's ethnic entrepreneurs* (1995) and *Shopping for Identity: the marketing of ethnicity* (2000).

Daniel Hiebert is Associate Professor of Geography at the University of British Columbia and a research coordinator at the Centre of Excellence for Research on Immigration and Integration in the Metropolis (RIIM) in Vancouver. His research focuses on several aspects of immigrant settlement in Canada, including the emerging social geography of immigrant neighbourhoods in Canadian cities, immigrant integration in the Canadian labour market, and the rise of immigrant entrepreneurship. Papers on all these topics can be found at the RIIM internet site (www.riim.metropolis.net), as well as in more traditional academic journals like the *Annals of the Association of American Geographers, The Canadian Geographer, Canadian Journal of Regional Science, Economic Geography* and *Progress in Planning*.

Trevor Jones was formerly Reader in Social Geography at Liverpool John Moores University. He continues part-time research at that university and is Visiting Professor at De Montfort University, Leicester. He studied at the London School of Economics and worked previously at Huddersfield Polytechnic. His collaboration with David McEvoy, Monder Ram and others on ethnic minority business and on ethnic segregation has produced papers in *Area, Annals of the Association of American Geographers, New Community, New Society, Social Forces, Sociological Review, Revue Européenne des Migrations Internationales, New Economy, Urban Studies, Work, Employment and Society,* and *Journal of Ethnic and Migration Studies*. Numerous essays in books have also appeared. He was senior author of *Geographical Issues in Western Europe* (1988) and *Social Geography: an introduction to contemporary issues* (1989).

Min-Jung Kwak is a PhD candidate in the Department of Geography, University of British Columbia, Vancouver. She received her MA in geography at York University, Toronto. She has been a research assistant in a number of Metropolis projects that investigated immigrant labour market integration in major Canadian cities. Her research has mainly examined the experiences of Korean Canadians, with special focuses on immigrant entrepreneurship, changes in gender and family relations, and transnational migration. She is preparing a doctoral thesis that explores the transnational entrepreneurial activities of Korean Canadians in the international education and tourism sectors.

Francisco Lima da Costa, a researcher on migration and ethnicity at SociNova, New University of Lisbon, also taught international tourism at the Polytechnic Institute of Tomar in Portugal. His research now focuses on the positive cognitive shift enabled by the market appropriation of immigrants' cultural expressions. He is also interested in ethnic tourism in migration contexts and its impacts on the political, socio-economic and cultural spheres of society.

M. Margarida Marques is a sociologist who has taught and researched on migration and ethnicity for the past decade at SociNova, New University of Lisbon. She led the first survey on immigrant entrepreneurs in Portugal in 2001, which generated several publications, including a chapter in a book on ethnic minority entrepreneurs edited by Leo Paul Dana (2005). Her recent research focuses on the economic and political spillover of immigrants' cultural production, on which the chapter in the present volume is the first contribution.

John C. Powers is a doctoral candidate at the Graduate School of Architecture, Planning and Preservation, Columbia University, New York. His research examines the process of innovation and techno-logical learning in locations undergoing rapid industrialization.

Monder Ram is Professor of Small Business and Director of the Centre for Research in Ethnic Minority Entrepreneurship (CREME), De Montfort University, Leicester. He is also Director of the Small Business and Enterprise Research Group at the same university. He has extensive experience of working in, researching and acting as a consult-ant to ethnic minority businesses. He has published widely on the subject of ethnic minority entrepreneurship. He is co-author of *Managing to Survive: Working lives in small firms and ethnic minorities in business* (1994), and has published papers in *Ethnic and Racial Studies; International Journal of Urban and Regional Research; Journal of Management Studies; Sociology; Entrepreneurship and Regional Development; Work, Employment and Society;* and *Journal of Ethnic and Migration Studies.*

Jan Rath is Professor of Urban Sociology and Director of the Institute for Migration and Ethnic Studies (IMES) at the University of Amsterdam (http://users.fmg.uva.nl/jrath). An anthropologist and urban studies specialist, he is the author, editor or co-editor of numerous articles, book chapters, reports and books on the sociology, politics and economics of post-migration processes. They include *Immigrant Businesses: the economic, political and social environment* (2000); *Unravelling the Rag Trade: Immigrant entrepreneurship in seven world cities,* with Robert Kloosterman (2002); and *Immigrant Entrepreneurs: Venturing abroad in the age of globalization* (2003).

Preface

More and more travellers, leisure seekers and business investors in gateway cities are indulging in ethnocultural events and festivals, and are gravitating to centres of immigrant ethnic commerce. The urban tourist economy is thus becoming one of the interfaces between immigrants from all strata of society and the wider economy. This tourism interface is an intriguing field of research.

Urban tourism has long been severely neglected as an area of academic research, but the body of literature has grown spectacularly in recent years. However, studies on how the urban tourist industry articulates with immigration or cultural diversity in advanced Western economies are still thin on the ground. Another rapidly expanding corpus informs us about the economic integration of immigrants in general and their labour market participation and entrepreneurship in particular. Yet those studies rarely if ever focus on the urban tourist industry. There is a clear need to explore the frontiers of knowledge on the interrelationship between tourism, migration, ethnic diversity and place, and to create new theoretical insights and new challenges for empirical research.

This book analyses the manifestations of ethnic diversity as commodified by immigrants and asks how these expressions of culture can be transformed into a vehicle for socio-economic development, to the advantage of both immigrants and the city at large. By addressing this problematic from an interdisciplinary and comparative perspective, this book seeks a more thorough understanding of the structural dynamics of immigrant activity in the tourist economy. The primary focus is on the role of immigrant entrepreneurs and workers in the emerging tourist industry and on their interaction with other players in the industry. The relative roles of public, private and civil society actors are important points of emphasis. Cases are drawn from several advanced economies in Europe, North America and Australia.

The book has evolved in part through discussions and collaborative work with a number of international researchers, including Marilyn Halter (Boston University), David McEvoy (Liverpool John Moores University), Maria Margarida Marques (Universidade Nova de Lisboa), Ching Lin Pang (Centre for Equal Opportunities and Opposition to Racism, Brussels), Monder Ram (De Montfort University), Martin Selby (Liverpool John Moores University); and particularly with Jock Collins (University of Technology Sydney) and Daniel Hiebert (University of British Columbia). I want to thank these scholars for their valuable and stimulating contributions.

I also want to express my gratitude to the Standing Committee for the Social Sciences of the European Science Foundation for awarding a grant for an Exploratory Workshop, and to the Social Sciences Internationalisation Fund (ISW) of the Netherlands Organisation for Scientific Research (NWO), and the Department of Economic Affairs of the City of Amsterdam, for co-sponsoring the preparations for this book.

Finally, I would like to thank Michael Dallas for his linguistic editing, and Heleen Ronden and Kim Jansen for completing the editorial process.

This book is dedicated to Mieke van de Rhee who, despite her untimely death, remains a fellow traveller for the rest of our lives.

Jan Rath
Amsterdam, the Netherlands

Acknowledgements

Acknowledgements of outside financial support

European Science Foundation, Strasbourg
Netherlands Organisation for Scientific Research (NWO), Social Sciences
    Internationalisation Fund (ISW), the Hague
Amsterdam City Council, Department of Economic Affairs
University of Amsterdam, Institute for Migration and Ethnic Studies
    (IMES), Amsterdam

1 Tourism, migration and place advantage in the global cultural economy

C. Michael Hall and Jan Rath

Tourism, migration and place are intricately linked. As early as the 1880s, it became fashionable for middle-class New Yorkers to go slumming or 'rubbernecking' in Chinatown (Lin 1998: 174), and in 1938 Vancouver 'officially' opened its Chinatown to tourism (Anderson 1988, 1995). In the 1970s, Melbourne courted the 'Chinese quarter' for its perceived distinctiveness and began sponsoring major redevelopment plans to boost such declining areas. Chinatown was selected as symbol of cultural diversity and object of civic pride and tourism (Anderson 1990). Many other places have followed suit, including San Francisco, whose Chinatown now ranks among the top five tourist attractions in a city where tourism is the number one industry.[1]

To be sure, these developments occurred in countries that had policies in place until the late twentieth century to keep their populations homogenously white and European. In the teeth of these exclusionary policies, the 'ethnic Others' gravitated towards those parts of the city where they could be amongst their own. Many such neighbourhoods were historically reputed as ghettos of the underprivileged and as places to avoid. San Francisco's Chinatown was one such district of ill fame. Around 1900, it was seen as a hotbed of crime and disease, a centre of drug dealing, a cesspool of vice, the home of criminals, hookers, vagabonds and rats (Denker 2003: 98–9). The tourist guides in those days strongly advised holidaymakers and travellers against visiting Chinatown. The distinguished Dutch scientist Hugo de Vries (1998: 215), who was staying in the Bay Area when the 1906 earthquake hit, wrote to his wife that San Francisco's Chinatown was badly damaged and that many of its residents were forced to leave. The professor laureate was jubilant: 'This is a great advantage!' The reputation of the Chinatowns of New York, Vancouver or elsewhere was not much better (Anderson 1995; Lin 1998; Wong 1995).

But, as always, times change. Numerous North American and Australian cities gradually came to take pride in their Chinatowns, Little Italys, Greektowns, Little Saigons, Little Havanas, Little Odessas, Punjabi Markets and all sorts of other ethnic precincts. Some districts now even take on the character of theme parks. (The chapters in this volume by

Alonso, by Collins, by Fainstein and Powers, and by Halter provide cases in point.) Travellers to the San Francisco Bay Area, New York, Vancouver, or Sydney will visit these neighbourhoods; they will stroll past the colourful specialty shops, try on unusual clothing, listen to world music, follow the exotic food trails, smell dozens of aromatic ingredients and enjoy *dim sum* or *pho* in one of the countless restaurants. Together, such ventures create a welcoming ambiance for international travellers who like to explore the 'world in one city' (Collins and Castillo 1998; see also the chapters by Collins, by Gotham and by Halter in this volume). Festivals are also abundant in precincts like these, and masses of people, including many from the local or tourist mainstream, enjoy and appreciate them greatly. The popularity of St Patrick's Day parades, Chinese New Year festivities, Puerto Rican Day parades, dragon boat regattas, ethnic food festivals and similar events is immense. The fact that these once-dodgy streets and neighbourhoods are now the sites of festivals and parades or the destinations of ethnic tours or city safaris, and that they are recommended in travel brochures, shopping guides and on the internet, and even incorporated into the city's place marketing campaigns, shows that this is not just a passing phase (cf. Collins and Castillo 1998; Morgenroth 2001; and other tourist guides). Ethnic diversity is cool!

Classic countries of settler immigration witnessed the emergence of a sort of ethnic tourism industry some time ago, and they have since acknowledged the tourist potential of ethnic diversity. In this respect they are ahead of Europe, but these types of tourist activities are now on the rise in Europe, too. European cities like Liverpool, Birmingham and Manchester have 'discovered' their old historic Chinatowns and Little Italys. Perhaps more importantly, in a diversity of precincts like Birmingham's Balti Quarter or London's Brick Lane ('Banglatown'), Belleville in Paris, Kreuzberg ('Klein-Türkei') in Berlin, Mouraria and Cova da Moura in Lisbon, or Dansaertstraat and surroundings in Brussels, immigrants are now carving out their own niches in the tourism industry – for example, by entering self-employment or by commodifying some of their cultural features (Shaw *et al.* 2004; see also the chapters in this volume by Aytar, by Jones and Ram and by Marques and Lima da Costa). Even in the Netherlands, where the political mood has abruptly turned against immigration and the concomitant cultural diversity, larger cities have shown a conspicuous eagerness to cultivate their 'Chinatowns' (Rath 2005). In the Hague, where Chinese immigration is only rather recent, the city government has actively promoted the transformation of Wagenstraat – a minor shopping strip along the 'exotic' City Mondial tour – into a Chinese precinct. Rotterdam and Amsterdam have had concentrations of Chinese immigrants since 1911, when Chinese sailors were recruited from London as strike-breakers and settled in their 'colonies' close to the docks (van Heek 1936). In Rotterdam, Chinese businesses have gradually moved out of the infamous Katendrecht district and gravitated to the

West Kruiskade area in the inner city, which city planners are now thinking of converting into a 'real' Chinatown with a gate, lions and all that jazz,
or at least into an 'exotic' shopping zone (de Gruiter 2000).[2] In Amsterdam,
Chinese businesses abound in the area around Zeedijk, Nieuwmarkt and
Geldersekade, inner-city streets adjacent to the famed red-light district,
another major tourist attraction. In the 1970s, these run-down streets were
the turf of street people and heroin addicts, but since the 1980s the
Amsterdam local authority has been revitalizing the area. It has tidied up
'skid row', renewed the streetscape, enabled the founding of the eye-
catching Fo Guang Shan Buddhist temple and interfaced with local business and community organizations. Zeedijk has gradually become a
magnet for people from all walks of life. With almost a million Google hits,
Amsterdam's Chinatown seems to be furthering the city government's goal
of branding Amsterdam as a centre of cosmopolitanism.

It is clear that immigrants in many countries – in this case by opening
and operating all kinds of tourist-oriented commercial ventures – are
providing what Australians would call a 'diversity dividend' or an 'ethnic
advantage'. Mainstream corporations have also become a party to this
development (Halter 2000; see also the chapters in this volume by Alonso,
by Fainstein and Powers and by Gotham). Interestingly, they have followed
the example of small, innovative ethnic businesses and entered the market
for cultural diversity. This demonstrates that the tourism industry, possibly
more so than most other parts of the knowledge economy, allows small
ethnic minority entrepreneurs to make an impact, even those who lack
specialized knowledge or substantial capital resources.

Tourism and migration – both defined in terms of mobility across
space – are integral to place competition. They are fundamental expressions of contemporary mobility, contributing to both the production and
the consumption of urban places. Urban centres are a focal point for the
tourism–migration relationship. They are the hubs through which human
mobility is channelled, as well as the primary space in which the vast
majority of immigrant settlement initially takes place. This function of the
city has been apparent in Western societies at least since the late seventeenth century and the onset of the Industrial Revolution, but arguably it
even dates from the period of European mercantile expansion in the
fifteenth century (Mumford 1973). In contemporary society, the role of
cities as the major 'staging post' in human mobility has gained added
significance by virtue of changes in transport technology and the accompanying development of new transport patterns – particularly the 'hub-
and-spoke' patterns that reinforce the dominance of some urban locations
and pre-existing transport networks. The development of new patterns of
accessibility are also intertwined with other processes of globalization,
identity and place formation – which, in their turn, have led to potentially
new patterns of consumption and production. Competition between
places has also meant that cities have sought to attract mobile populations

(of the 'right' kind) as well as mobile capital. Indeed, just as places in the nineteenth century competed for the railway and places in the twentieth century for the motorway, so cities now jockey for air connections, high-speed train stations or cruise terminals – all in order to ensure continuing links to the global cultural and economic network, and thus to attract and (for a while at least) hold onto the mobile.

Enormous competition between locations takes place for highly skilled labour, for immigrants with capital and, in many cases, for lower-paid service sector workers. Tourism is important not least because of the perceived direct economic value of tourist spending. But it is also significant for the 'halo effects' that tourism marketing can have on broader place imaging and promotion, and for the role that tourist visits might even play as a precursor to longer-term settlement.

The tourism industry is one of the fastest growing sectors of today's world economy. In some countries, tourism's contribution to the gross domestic product (GDP) has surpassed that of any manufacturing sector. According to the World Travel and Tourism Council (2005), travel and tourism were expected to generate US $6.2 trillion (that is approximately €5.2 trillion) of economic activity in 2005, and to grow to $10.7 trillion (€8.8 trillion) by 2015. Over the entire world today, they contribute an average of 3.8 per cent to GDP and are responsible for an estimated 74,223,000 jobs, or 2.8 per cent of the world's work force. Moreover, since travel and tourism touch all sectors of the economy, their real direct and indirect impact accounts for up to 10.6 per cent of total GDP, or 221,568,000 jobs. That represents 8.3 per cent of total employment. A large part of the tourism and travel economy is generated in the European Union, and by 2011 Europe's share is expected to grow to 32 per cent of the global total per annum in real terms. This varies among regions and countries, of course, but the gist of the argument remains the same: tourism is and will continue to be perceived by many as a job machine.

As has often been observed, tourist enterprises run by immigrants are increasingly attracting people with resources to their cities and communities (Fainstein and Judd 1999; Hoffman 2003a, 2003b; Rath 2005; Shaw *et al.* 2004; and the chapters by Bartling, by Gotham and by Kwak and Hiebert in this volume). Few attempts have been made to measure the size of the immigrant tourist sector. Prescott and Wilton (2000) investigated the relationship between immigrant landings from different countries into Canada and the flows of tourists from those countries. They found that each immigrant to Canada arriving between 1988 and 1997 generated C$4500 (€2500) of spending by foreign visitors. If immigrant intake had been reduced by 50 per cent, C$200 million less would have been spent by foreign visitors to Canada. It is safe to assume that much of this money would be spent in minority ethnic economies (at ethnic travel agents, ethnic restaurants and other such enterprises).

Today it is hard to imagine urban tourism without immigrants. Immigrants often provide the cheap labour or the entrepreneurial drive for the urban tourism industry. Moreover, their real or imagined cultural expressions are also increasingly discernible amongst the 'objects' of urban tourism. The involvement of immigrants in the tourism economy assumes two general forms.

- Many immigrants work as *low-skilled employees* for mainstream tourism-related companies, for instance as waiters, cooks, dishwashers, valets, hotel porters, security men, pool attendants, chambermaids and janitors. As Adler and Adler (2004) and Aytar and Bartling (in their chapters in this volume) convincingly demonstrate, this employment is sometimes stratified along ethnic, racial and gender lines – according to real or ascribed expectations on the part of tourists, skill levels, the quality and remuneration of jobs, and whether or not jobs are seasonal. Many low-skilled workers toil in unfavourable conditions with little prospect of upward social mobility. But, as the above authors also show, some jobs in the tourist sector practically amount to tourist consumption themselves, as seasonal migrant workers (chiefly white and Western) indulge themselves while working in cool clubs, famous holiday resorts or popular amusement parks.
- Immigrants are also involved in the tourism economy as *self-employed entrepreneurs*. They pop up as owners of cafés, coffee shops, restaurants, travel bureaus, hotels, souvenir shops, telephone and internet shops, or as organizers of cultural events and as web designers. It should be noted that not every immigrant entrepreneur ventures into business with the *intention* of establishing mass tourist attractions (although some definitely do). Most simply set up shop with a view to becoming self-sufficient and achieving upward social mobility. These fledgling entrepreneurs often gravitate to markets with lower entry barriers. Their businesses may nonetheless contribute to the 'ethnic atmosphere' of an area, and in so doing they may help foster the ethnic tourism economy. In the same vein, immigrant entrepreneurs did not always start organizing festivals and cultural events with the intention of attracting a wider public. Often such activities were part of an ethnic community-building process (Gotham 2002; Knecht and Soysal 2005; Shaw *et al.* 2004). Immigrants were seeking ways to interface with other people from the same group, or to make the past in their country of origin live again, or to assert their ethnic identity vis-á-vis the rest of society, oftentimes without even intending to convert these events into money-making endeavours. Hence, the experiences, intentions and perceptions of the organizers do not necessarily coincide with those of the outside visitors who come to seek exoticism, authenticity or just an afternoon of pleasure. That said, it is clear that the presence of hundreds, thousands and sometimes hundreds of thousands of

spending spectators can have a far-reaching impact on the immigrant business community. It therefore comes as no surprise that immigrant entrepreneurs have increasingly taken to organizing such events.

Tourism, migration, ethnic diversity and place are structurally related in many ways, but this interrelationship has been surprisingly understudied. Urban tourism as a whole has been a severely neglected area of academic research (Page and Hall 2003), despite its economic and social significance, as shown by Ashworth and Voogd (1990) years ago. Many studies on international tourism have focused on areas far removed from the centres of industrial power, such as island societies, territorial enclaves, ecotourism in the hinterland and hiking trails; some other studies have dealt with heritage tourism – the metamorphosis of historical places into tourist attractions. The body of literature on mainstream urban tourism has somewhat expanded in recent years (some examples are Hoffman *et al.* 2003; Judd and Fainstein 1999; Page and Hall 2003; Selby 2004). However, the numerous ways in which that industry articulates with *immigration* or *cultural diversity* in advanced Western economies is still largely ignored (with Gotham 2002; Hall and Williams 2002; Hoffman 2003a, 2003b; Rath 2005; Shaw *et al.* 2004 and perhaps also Frenkel and Walton 2000 as notable exceptions).

An entirely different body of literature addresses the integration of immigrants into the urban economy, but it rarely if ever focuses on the urban tourism industry. Most studies about Chinatowns, Little Indias and other ethnic enclaves (Zhou 1992) tend to examine the internal dynamics only, neglecting the roles of wider political, economic and cultural processes, and barely mentioning the potential for tourism (exceptions are Conforti 1996; Lin 1998; Timothy 2002 and a few others). A clear need therefore exists to explore the frontiers of knowledge on the interrelationship between tourism, migration, ethnic diversity and place, and to create new theoretical insights and new challenges for empirical research.

This book deals squarely with that relationship. It explores the manifestations of ethnic diversity that have been commodified by immigrants in gateway cities, and it asks how these expressions of culture can be transformed into vehicles for further developing the urban tourism economy. Our primary focus is on the role of immigrant entrepreneurs and workers in the emerging urban tourism industry and on their interactions with other players in that industry. The relative roles of public, private and civil society actors are important points of attention. By addressing these issues from an interdisciplinary and comparative perspective, we seek a more thorough understanding of the structural dynamics of immigrants' commercial manifestations of ethnic diversity. We further examine how such activities serve to integrate immigrants into the knowledge economy and how they impact upon urban socio-economic development as a whole.

This set of issues ties in with several different bodies of research. One of these is the literature on globalization and the rise of post-industrial, service-oriented urban economies. Another investigates the emergence of 'creative cities' and the growing significance of cities as places of consumption. Still another deals with immigrant or ethnic entrepreneurship itself. In the remainder of this introductory chapter, we explore how urban economies in general have become transformed in the past few decades, and we ask what impact these fundamental changes have had on the socioeconomic opportunities of native and immigrant city dwellers. We briefly discuss the role of immigrants as entrepreneurs and how such forms of economic involvement are linked to wider social, political and economic structures. We examine the process of revaluation of urban space and, more specifically, the commodification of culture as a key element in that process. We conclude by addressing several factors that shape the ways in which immigrants are involved in the urban tourism industry.

Globalization and the rise of post-industrial service-oriented urban economies

In the past few decades – roughly coinciding with the rapid increases in international migration – the advanced economies of Europe, North America and Australia and New Zealand have undergone fundamental changes. These transformations were heralded by a steady decline in the agricultural and manufacturing sectors and an unprecedented loss of industrial jobs. They were associated at the same time with a spectacular growth in the service and knowledge industries. ICT, finance, insurance, real estate, cultural industries, media and tourism have all developed into major sources of income and economic growth, offering opportunities to millions of different kinds of people (Castells 1996). The expansion of personal and producer services has been fuelled by a wide array of inter-related factors, including a fragmentation of consumer tastes, the introduction of ever newer information technologies, the erosion of economies of scale, changes in the international division of labour, an increased mobility of capital and labour, the diminishing role of the state as a comprehensive provider of services, and a concomitant reappraisal of the private sector. Whatever factors may have been decisive, the economy that has emerged in these 'Western' countries is now largely deindustrialized and global, and places a high premium on knowledge and information professionals (Florida 2002; Landry 2000; Scott 2006).

While not wishing to argue that any one factor can be isolated from the others, we would note that economic development in advanced regions has long been associated with international migration, even if the nature and function of immigration has changed over time. Northern and Western European countries, for instance, were already receiving immigrants from former colonial areas, as well as predominantly male, unskilled workers from Mediterranean regions, from the 1950s up to the mid-1970s – that is, before the forces of economic restructuring gained momentum. The Mediterranean workers were recruited under 'guest worker' schemes and, along with many of the ex-colonials, were allocated to low-wage 'dirty, dangerous, and difficult' (3D) positions in the labour market. Mainstream workers who had opportunities to move up the social ladder were vacating these jobs. That was especially the case in the 'old' manufacturing industries such as textiles, garments, ship-building, metal, shoemaking and meatpacking. International migrants played a critical role at this juncture. Industrial restructuring was already an emerging phenomenon, but the employment of low-wage migrant workers enabled the sunset industries to temporarily sustain their production and their internationally competitive positions. This allowed them to gradually adapt to the new political and economic realities of the knowledge-based society.

The evidence is, however, that unskilled people find it harder to benefit from the knowledge economy. They experience relatively high levels of unemployment, even in times of boom, and low levels of upward occupational mobility. This is particularly the case for immigrants from Third World countries. If they are active in the labour market, they tend to populate sectors with a high demand for manual or unskilled workers and with low entry barriers, such as cleaning or catering (Engelen 2001; Rath 2002). This indicates that there is a growing, and ethnically specific, divide between the highly educated, well-paid knowledge workers in Western societies and the workers that are concentrated in the lower tiers of the labour market or even more seriously marginalized. These developments are also gender specific, as the changing opportunity structure produces different outcomes for men and for women.

These crucial trends also have their own spatial dimension. It is argued that capital today is more mobile, and that cities are even more central to the emerging economy, than they were in the Industrial Age (Castells 1996; Webster 2001). Gateway cities like Miami, New York, Vancouver, Sydney or Amsterdam have become nodes in national and international networks, bridging migrant communities and linking businesses and consumers from all over the world. High-skilled professionals, low-skilled job seekers, students and holiday travellers alike are gravitating there. They are part and parcel of the rapid transformation of these cities from sites of industrial production into spaces of information circulation and consumption. Kwak and Hiebert (this volume) show how these multifarious categories of migration are intricately linked together: the rise of the international education industry initially involves the (officially temporary) migration of students, but it also generates new forms of international mobility, including visits by family members and friends to the international students and the arrival of business people who cater to the needs of these categories.

City authorities show interest in these developments, if only because they believe that the students of today are the international job seekers or business investors of tomorrow.

These converging processes affect the urban social fabric in profound ways. They alter the opportunity structures, shape and reshape forms of inclusion and exclusion, and add new dimensions to the already existing economic, social and cultural diversity. Sociologists and geographers like Castells, Fainstein and Sassen have captured this socioeconomic and spatial bifurcation, arguing that the dual city is at once globally connected and locally disconnected (Castells 1989; Fainstein *et al.* 1992; Mollenkopf and Castells 1991; Sassen 1991). This social division inhibits urban socioeconomic development and undermines the quality of urban life. Clearly, those who fail educationally and are excluded from the knowledge economy, yet remain stuck in city areas in close proximity to the rich and affluent, become a matter of serious political and social concern. In these circumstances, the rise of urban tourism, and the new opportunities it creates in the service industries, is more than welcome.

Commodification of culture and revaluation of urban space

Here we enter the realm of studies on the emerging knowledge economy and the role of cities as places of information circulation, creativity and consumption. The growth of the urban tourism industry is intricately tied up with the rapid transformation of the manufacturing economy to an information economy and beyond. Deindustrialization resulted in a need for localities to differentiate themselves, in order to attract a share of the spatially mobile capital. In large cities in particular, authorities ranging from local governments to marketing consortia now strive to present their localities as attractive places for potential investors, employers, inhabitants and tourists (Kearns and Philo 1993).

Urban tourism is more than just a collection of tourist facilities. It is the consumption of signs, symbols and spectacle, the experiencing of aestheticized spaces of entertainment and pleasure (Featherstone 1991; Kearns and Philo 1993; Lash and Urry 1994; Selby 2004). The manipulation of place images and the projection of a high quality of life reflect a revaluation of urban space at the local level in response to global processes. The deliberate creation of an attractive place product and place image, together with the material processes that produce the urban landscape, spawns a diversity of spatial narratives (Duncan and Duncan 1988; Zukin 1991, 1995). The existence of spatial narratives and their contested nature are central to debates on the utilization of tourism, place marketing and heritage in socioeconomic development.

By pursuing urban imaging strategies, cities seek to attract mobile capital, tourists and immigrants (so long as these fit into the prescribed urban policy goals). As Roche (1994, 2000) has noted, the principal aims of imaging strategies, or re-imaging in the case of older urban centres, are to:

- attract visitor expenditure
- generate employment in tourism and other industries
- foster positive images for potential investors in the region, often by 're-imaging' previous negative perceptions
- provide an urban environment that will attract and retain the interest of professionals and white-collar workers, particularly in 'clean' service industries such as tourism and communications.

These strategies, in short, are employed to attract mobile capital and people.

Urban imaging processes are typically characterized by some or all of the following activities:

- creation of a critical mass of visitor attractions and facilities, including new buildings and prestige centres (casinos, event and convention centres, stadiums, refurbished waterfronts);
- the hosting of hallmark events;
- development of urban tourism strategies, policies and campaigns, often associated with new or renewed organization and development of city marketing;
- development of leisure and cultural services and projects to support the marketing and tourism effort (e.g. the creation or renewal of museums and art galleries and the hosting of art festivals, often as part of a comprehensive cultural tourism strategy for a region or city);
- deliberate commodification and rebranding of urban space to identify 'new' places for leisure consumption (e.g. the promotion of nightlife, restaurant or ethnic districts).

Whilst urban imaging processes help to explain why some areas or activities have become so popular, we also need to explore how these patterns of consumption are shaped. Why have deprived neighbourhoods in cities like Amsterdam, Berlin, Birmingham, Brussels and Lisbon turned into magnets attracting numerous day-trippers who want to become absorbed into an exotic world (cf. Collins and Castillo 1998)? Why has the mainstream diet in countries like the Netherlands, Germany and England, which centred around bland meat, potatoes and standard vegetables, expanded to embrace tacos, Bombay duck, chicken tikka masala, döner kebap, rijsttafel and babi pangang (Halter 2000; Miller *et al.* 1998; Narayan 1995; Ram *et al.* 2000; Tsu 1999)? What cultural processes support the growing popularity of 'ethnically' diverse attractions by persuading people to appreciate new and different tastes?

These processes, to be sure, do not just arise from the manipulations of local government officials. Zukin (1991; and see also Valle and Torres 2000) has argued that a *critical infrastructure* of individuals must exist who are all connected in some way with cultural production and appreciation and who are capable of influencing public taste. This critical infrastructure contains a broad spectrum of knowledge workers that design cultural production and consumption; they include connoisseurs, cultural mediators and marketing bureaus as well as business associations, tourist boards and parts of national and local government. Whatever the intentions behind it, the critical infrastructure influences the popularity of particular cultural products. It identifies which products are 'relevant' and makes statements concerning their real or alleged authenticity and the significance of consuming them. It thereby helps to shape the multicultural urban landscape. The messages from this critical infrastructure are not simply 'free-floating', but are conveyed through the internet or more conventional means of communication, through the behaviour of trendsetters or other people from the cultural vanguard, through the programmes of specific educational institutes, or through the ready availability of easy-to-consume products and services. As Zukin (1991) points out, shifting preferences have an impact on the nature of production; they ultimately bestow opportunities on some groups and their areas of the city, while simultaneously making other groups and areas largely invisible to 'the people who matter'. How this critical infrastructure functions, and how immigrant producers of tourist attractions, as well as their consumers, fit into this process, are a matter for further research. The chapters in this volume by Fainstein and Powers and by Marques and Lima da Costa provide good evidence of the operation of such infrastructures.

Zoned ethnicity

Identifiable ethnic areas within cities are now often used as an attraction for short- and long-term visitors, thus contributing to the identities of the cities themselves, of those who live in the areas and of those who visit. As observed above, few researchers have studied ethnic enclaves, or the 'zoned ethnicity' of some parts of cities, in terms of their tourist dimensions, even though many of them are already notable urban attractions (Abrahamson 1995; Timothy 2002).

Let's return to the case of San Francisco's Chinatown. The original Chinatown was widely considered a slum, and had a bad reputation in white American society as the dumping ground for an undesirable immigrant group. However, as Timothy (2002; cf. Anderson 1995; Lin 1998; Wong 1995) has observed, this very image paradoxically encouraged tourism development. To non-Chinese people in the early twentieth century, Chinatown was a quaint, mysterious part of town, a 'foreign colony' in their midst. Advertisements promised that white tourists visiting the district would experience the 'sounds, the sights, and the smells of Canton'. They could imagine themselves in 'some hoary Mongolian city in the distant land of Cathay', and could 'wander in the midst of the Orient' without leaving America, seeing throngs of people with 'strange faces' in the streets eating 'chop suey'. Over time, the San Francisco city authorities began to commodify Chinese ethnicity and promote Chinatown as an exotic Chinese colony; its back alleyways were presented as picturesque little lanes for the purposes of visitor consumption. Tours began running which took visitors through the crooked, narrow streets of Chinatown past souvenir shops, stores and restaurants. White Americans viewed Chinese enclaves as alien districts in the heart of America – as curiosities to visit. In the early 1900s, tourism helped Chinese Americans to survive after racial discrimination had prevented them from competing in the main-stream labour market. Around the world, Chinatowns are still significant tourist destinations to this day, although social and structural conditions there have improved immensely and the original social and political func-tions of such ghettoes have changed. The Chinese enclave in San Francisco, for instance, home to about 30,000 Chinese Americans, is now one of the city's most popular attractions.

Italian ethnic islands have also furnished much of the urban lifestyle milieu that appeals to some tourists, particularly in North America and Australia (Fremantle, Melbourne). Commenting on the popularity of urban Italian communities, Conforti (1996: 831) observes that

> Little Italies are familiar to many American tourists (and foreign visitors), usually as somewhat exotic and alien places that are quasi-foreign, where interesting food can be found, exotic people can be observed, and even a lurking danger (as the home of the Mafia) can be sensed.

He argues that tourism is a driving force for keeping the image of Little Italy alive in cities like Baltimore, Boston and New York (see also the chap-ter by Halter in this volume), making it 'a symbolic area with a pronounced and entrenched image shared by insiders and outsiders, an image based on the area's history and use, especially as a tourism attraction'. Indeed, as Collins (this volume) and Timothy (2002) observe, in some cases tourism drives the efforts to preserve Little Italys as distinct ethnic enclaves even after all the local Italians have gone. Yet the importance of ethnic clustering in urban regions lies not just in a historical context; for many ethnic groups today, inner cities still house their commercial centres, and these older migration destinations are often still the site of first settlement for new and successive waves of migration in the same groups.

Given the interrelationship between identity, image and urban place promotion, it is not surprising that the links between particular ethnic and cultural groups ('the Other') and particular places have been commodified for the purposes of place promotion. This process can assume various forms, ranging from publicizing the availability of national and regional cuisines to the commodification of entire localities, such as Chinatown, or of entire ethnic or national identities. As Jules-Rosette (1994) has shown with respect to Afro-Antillean Paris, the transformation of locales of every-day life into tourist attractions linked to the identity of a particular foreign ethnic population is part of the process of *postmodern simulation* in tourism. Cultural identities may be explicitly utilized in the branding of places on large scales, e.g. Scotland (McCrone *et al.* 1995); they may also be integrated into tourism policy to further domestic political aims, as has happened in Singapore (Chang 1997). Yet the role of immigrant communi-ties in such situations is actually problematic, because the use of the commodified ethnicities of immigrant communities is dependent both on contemporary political realities and sensitivities and on the extent of linkage between the local and the global. Without such a linkage, no commodification of ethnic space will be possible in the service of the place-marketing and exchange-value objectives of local growth coalitions. In some places, ethnicity is commodified as a heritage product only (with contemporary ethnic realities ignored); in others, commodification focuses on more recent immigrant settlement; while in many locations the immi-grant Other is completely ignored as tourist product. Indeed, if we consider the numerous advantages that immigrant populations could potentially have as tourist attractions, we might well conclude that this potential is currently not utilized anywhere near as much as it could be for place promotion and imaging. So the creation of commodified 'ethnic precincts' for the purposes of tourism needs to be understood in terms of the links between the local and the global, rather than simply in terms of municipal actions (see the chapter in this volume by Collins), however essential the latter may be. In tourism terms, the commodification of ethnicity for tourism and promotional benefits is articulated into the national and inter-national tourism industry through distribution and promotion channels. Such articulation is openly supported, if not actively developed, by city and national growth coalitions – sometimes even without the support of the ethnic groups that are being commodified as a place product.

As suggested above, the diasporic relationships and transnational networks of immigrant communities generate a significant proportion of tourism-related travel in their own right, as well as providing a foundation for the development of domestic and international business networks. They act as carriers of information and knowledge, provide mutual aid and assistance, and encourage entrepreneurship from within the ethnic community (Thrift and Olds 1996; Yeung 2002; Yeung and Olds 2000). Overseas Chinese firms, in particular, have been shown to draw heavily on such social and economic networks in the Chinese diaspora, but parallels have been found in many other transnational diasporic communities as well (Coles and Timothy 2004; Lew and Wong 2002). Tourism is a vital component of such transnational networks, as travel helps to maintain family, personal and business ties. Indeed, the promotion of multicultural policies in recent years with respect to immigrant populations in some countries has likely served to reinforce such transnational networks, as transnational spaces continue developing within diasporic cities (Cartier and Lew 2004; Zukin 1995).

Mixed embeddedness

Tourist services, particularly in the hotel and restaurant sector, often rely on new immigrant communities to fill low-skill, low-wage service jobs that increase business viability. Where tourism development outstrips the local supply of labour, that will lead to in-migration. Such mobility into tourism is usually facilitated by the low barriers to entry in terms of the skills and qualifications required for most jobs, as well as by the sheer range of jobs. To some extent, of course, the contingent nature of the local labour market – especially the levels of job availability and wages outside the tourism sector – is an important factor. In-migration is most likely to surge under conditions of mass or urban tourism and leisure development; smaller-scale rural tourism may be able to satisfy most labour demands from within the local labour market. Some types of tourism and hospitality may require specialist skills, such as those of qualified chefs, and this may entail recruitment from outside local or national labour markets, often through temporary work visas. Similarly, the recruitment of staff to serve particular foreign language markets may also involve attracting either short-term staff with specialized language skills or even migrants with the needed language backgrounds, as when Australia and New Zealand recruited Japanese and Korean speakers in the 1980s and 1990s (Adler and Adler 2004; Hall 2007).

This draws our attention to the entrepreneurial side of tourism development. Most studies of immigrant or ethnic entrepreneurship focus on the business people only, trying mainly to explain the propensities of certain groups towards entrepreneurship and to describe their paths to entrepreneurial success.[3] Several theoretical approaches have emerged, some emphasizing the cultural endowments of immigrants (such as a cultural inclination in certain groups towards risk-taking behaviour), some highlighting racist or ethnic exclusion and blocked mobility in the mainstream labour market (as when marginalized individuals are driven into entrepreneurialism as a means of escaping unwelcoming labour markets; for an overview, see Kloosterman and Rath 2003). Other approaches revolve around issues of social embeddedness. They argue that individual entrepreneurs take part in ethnically specific economic networks that facilitate their business operations (especially when it comes to acquiring knowledge, distributing information, recruiting capital and labour, and establishing relations with clients and suppliers). Social embeddedness enables entrepreneurs to reduce transaction costs by eliminating formal contracts, thus gaining privileged access to vital economic resources and ensuring safeguards against misconduct. Particularly in cases where the entrepreneurs' primary input is cheap and flexible labour, as is true in some parts of the tourism industry, this cost reduction by mobilizing social networks for labour recruitment seems crucial. Yet drawing benefit from social embeddedness is a complex and dynamic process. It involves cultural, human and financial capital (Light and Gold 2000); it is contingent on the goals pursued and the political and economic forces at work (Granovetter 1995; Kumcu 2001); and it depends on the interaction of structural factors such as migration history and processes of social, economic and political incorporation into mainstream society. All such influences are also subject to spatial variations.

In recent years, continental European researchers have been exploring these complex interactions along with the array of regulatory structures that promote certain economic activities while inhibiting others (Engelen 2001; Jones *et al.* 2000; Kloosterman and Rath 2003; Ram 1998; Rath 2000, 2002). Although the latter forms of regulation are highly relevant, researchers argue that it is important to fully understand the economic dynamics of a market. It does not require much sociological imagination to see that souvenir vendors, take-away restaurateurs or designers of virtual tourist guides operate in entirely different markets. Different markets imply different opportunities and obstacles, require different skills and lead to different outcomes in terms of business success or – at a higher level of agglomeration – to different ethnic divisions of labour. The situation is also complicated by differences between stated and observed behaviour in firms with respect to ethnic or migrant skills. As Aitken and Hall (2000) have noted in the case of New Zealand, although potential employers in the tourism industry insisted that migrant skills were valued in the hiring of staff, the extent of the actual employment of migrants did not match the rhetoric, particularly at the management level.

Acknowledging the importance of both regulation and market dynamics, researchers have proposed a 'mixed embeddedness' approach to immigrant entrepreneurship (Kloosterman and Rath 2001, 2003; Rath 2000, 2002). This approach is more encompassing in that it links social relations and transactions to wider political and economic structures. It acknowledges the significance of immigrants' concrete embeddedness in social networks while understanding that their relations and transactions are more abstractly embedded in wider economic and political-institutional structures.

This approach also devotes more attention to the demand side – the consumers – an aspect often overlooked in the entrepreneurship literature. Obviously, tourist entrepreneurs, or any other business people for that matter, cannot survive without a demand for their goods and services, and it is therefore necessary to examine the critical role played by consumers and by their interactions with entrepreneurs. Interaction between producers and consumers of tourist attractions is not, of course, confined to one-to-one relationships. In the tourist market, as elsewhere, supply and demand are mediated by other factors.

Facilitating the use of migrant identity in urban place promotion

We therefore need to be aware of the key factors that facilitate the use of migrant identity in place promotion. A number of interrelated conditions seem necessary:

- *Growth coalitions.* Local growth coalitions have emerged in many communities, based on a mixture of coercion and cooptation aimed at maintaining real estate values, assumptions regarding the generation of employment and investment, and a belief that growth will automatically be beneficial. Such coalitions, often linked to activities of certain local business champions (usually with strong connections to City Hall), are an essential starting point for the commodification of ethnic diversity to promote tourist and lifestyle consumption. Growth coalitions, sometimes described in terms of 'civic boosterism', tend to be enthusiastic supporters of place-branding mechanisms (whether or not their local communities actually *want* to be branded). They regard place branding as good for business and therefore good for place, and they usually assume it raises the exchange value of properties in the commodified space. Ethnic tourism activities must, therefore, be part of a larger tourism industry. If such a larger industry does not really exist in a city, the odds are against a quick launch of an ethnic tourism sector. The 'difficult area' of Bradford in England – 'not a "natural" tourist attraction' – is a case in point. A local government strategy to attract leisure tourists, partly by promoting the 'flavours of Asia' and by garnering funds from the private sector and the European Union, has not fundamentally improved the attractiveness of the city since the mid-1980s (Hope and Klemm 2001: 633). Although some improvements were made to the infrastructure, the image of Bradford remained unfavourable.
- *Political regulation and structure.* The existence of growth coalitions is one thing, the formal control of political structures and institutions or the creation of new ones is another. In many urban centres in the Western world, the systems of governance and the role of the local state have changed markedly in recent years, but this transformation of policy settings and its implications for place promotion have not yet been well documented. In any case we can observe that the urban local state is increasingly interested in the symbolic economy, and this resonates in its approach to immigrant ethnic minorities. Local policies

today – to the extent that they are not couched in revanchist discourses or geared to assimilating or even disciplining minority groups – often acknowledge that immigrants provide business networks and access to international markets, as well as having value for attracting tourists and creating 'vibrant local culture' (Zukin 1995). The language of cultural policy therefore focuses increasingly on the commodity values of culture as an element of place advantage. 'Elected officials who, in the 1960s, might have criticized immigrants and non-traditional living arrangements, now consciously market the city's diverse opportunities for cultural consumption' (Zukin 1998: 836). This process amounts to a commodification of diversity, and it has created conditions in which culture – immigrant and minority cultures in particular – can be under-stood as an economic resource for cities. In this light, the policy issue for the urban local state is now often how to maximize the advantage derived from the presence of visible immigrant communities, or how to enhance their asset value in terms of urban and regional development.

- *Spatial confinement*. Although strategic linkage to international business networks, international business skills and new markets is one dimension of the desire of growth coalitions to incorporate some immigrant communities from a place promotion perspective, such measures may not be sufficient to attract large numbers of tourists from *outside* the social networks of the immigrant groups. For this, immigrant culture needs to be commodified. This can take place in many ways, such as hosting events and festivals or supporting artistic groups. Arguably the most attractive option for tourism promotion is to ensure that certain locations or regions have an identifiable Otherness associated with specific immigrant groups. This requires at least a substantial immigrant presence in inner city areas or in large suburban hubs, as well as a 'shop-front' immigrant visibility there. Such spatially identifiable areas may have historical associations with immigrant populations, as with the Chinese ghettos later re-branded as Chinatowns. More often than not, planning regulations will be introduced that encourage and sustain such zones of Otherness over time. There is, however, a risk that such locations will degenerate into fossilized urban landscapes, depleted of economic dynamics (Anderson 1995; Lin 1998).

- *Immigrant entrepreneurship*. The mere presence of large numbers of immigrants from certain groups in suburban areas does not automati-cally constitute a visitor attraction. A high street or several street blocks are needed which radiate characteristics associated with those populations. There should be a proliferation of shops that lend the neighbourhood its ethnic flavour and stimulate street life, giving tourists an excuse to hang around and pass the time. Book and music stores, gift shops, travel agents, and especially restaurants, grocery shops and ethnic supermarkets have such qualities.

- *Ethnic infrastructure.* A social infrastructure must be present that can support the transformation of an ordinary precinct into a tourist attraction. Ethnic precincts – except perhaps the manufactured ones such as the Chinatown in Las Vegas or the China Pavilion in Walt Disney's theme park Epcot in Orlando (see the chapter in this volume by Bartling) – are commonly the product of the activities of immigrant ethnic communities. There should be a space that serves as the nodal point of community life. The ethnic community members do not actually have to live in the precinct. Sydney's Little Italy, Leichhardt, is no longer the home of the Italian population, but it is still the place where the Italian community congregates and where the sights, sounds, flavours and irresistible aromas of Italy come alive (Collins and Castillo 1998; see also Halter's contribution to this volume). Neighbourhoods like these are commonly perceived as 'authentic' (whatever that may mean). An essential related issue is the immigrant group's acquiescence in the process of tourist commodification, as conflict is anathema to tourism development. It often happens that the 'local colour' lent to a neighbourhood by its immigrant restaurants, shops and inhabitants is initially put to use in tourism promotion activities without the local community immediately being aware of what is happening. For any longer-term commodification and promotion to succeed, however, immigrant communities will need to be supportive. This is usually achieved through organizations and business people from the immigrant group itself that have become a part of the urban growth coalition (often this commodification process is highly gendered). While some resistance may occur, business people from the immigrant groups usually see such place promotion as a business opportunity, rather than as a loss of culture. Stereotypes about the 'authentic' ethnic Other may even be *reinforced* (Anderson 1990; Fainstein and Gladstone 1999; Mitchell 1993). The issue is whether immigrants, as the 'toured Other', are being reduced to tourist objects or, quite to the contrary, are emerging here as active subjects in a new tourist practice in the city (Morris 1995). Lin (1998: 205), while acknowledging that the tourist industry entails a 'performative repertoire of cultural displays that increasingly serve the consumptive and spectating demands of outsider audiences', also points to the risks of such voyeurism and stereotyping. Immigrants may interpret their exposure to an ever wider public as undesirable interference with their own affairs, or even as a kind of cultural imperialism.
- *Accessibility.* Spatial confinement, though necessary for the production of tourist space, is only one side of the commodification process. Ethnic tourism neighbourhoods also need to be accessible to visitors. If they are not, there will be too many other attractions or alternative locations to compete with for tourist consumption (Hall 2005).

- *Safety*. There is a critical connection between the prevalence of crime, government programmes to combat crime and the development of urban tourism. The prevalence of crime in ethnic commercial precincts may undermine their prospects of becoming tourist attractions.
- *Target marketing*. All of the above factors pertaining to ethnic identity and place promotion are contingent on an underlying assumption: the immigrant group in question must truly be considered by place promoters and other members of the critical infrastructure to be an attractive element for market activities. As described above, this is the broad critical infrastructure of knowledge workers that shape and promote tourism attractions by defining 'relevant' cultural products, disseminating information on them and ensuring their availability. This does not just happen as a matter of course. Certain ethnic concentrations in some cities may be attractive to overseas markets but not to the local market, or vice versa. Auckland in New Zealand, for example, promotes itself for the purposes of international tourism as the 'world's largest Polynesian city', but it does not do so within New Zealand. Arguably, this is not only because New Zealanders may not recognize the attractive Otherness of Polynesians, but also because the issue may come too close to national debates on identity and race. On the other hand, immigrant locations, usually associated with food kiosks, restaurants and shopping opportunities, may be extensively promoted on some domestic markets. At the risk of generalizing from the fieldwork undertaken by the first author, it is noticeable how the commodification of immigrant groups in the place promotion of Western cities has increasingly shifted from identifiable European populations to Asian and Caribbean populations. This seems clearly related to altered flows of migration and to intergenerational change, but also, and perhaps more strongly, to the altered patterns of mobility in the Western world (Hall 2005). Westerners with the mobile lifestyles we often associate with tourism, and who are also the main *in situ* consumers of leisure space, have meanwhile become so 'familiar' with Italy that they no longer need a Little Italy. Such neighbourhoods have ceased to convey a sufficient vicarious Otherness to be attractive to that group. Nonetheless, as the attraction to domestic consumers wanes, the neighbourhoods may ironically be transformed into places of Otherness for new sets of international visitors.

Conclusions

The tourism industry is one of the fastest growing sectors of the new service economy, providing opportunities to immigrants of both genders. In principle, the industry can form a powerful interface between skilled or unskilled immigrants and the wider knowledge economy. When the sector fulfils its promise and enables the commodification of immigrants' ethnocultural resources, it contributes to the making of the cosmopolitan city, thus enhancing the city's potential to attract domestic and international knowledge workers and business investors. Equally important, it strengthens the social and economic integration of immigrants. This is not an automatic chain of developments, however, and even if it were, not every immigrant would automatically achieve upward social mobility or contribute to the full acceptance and integration of immigrant communities.

The promotion of new urban images, new lifestyles and new 'city myths' is often a necessary prelude to the establishment of new urban economies. Immigrant communities are a vital element of the new urban economy in many ways. Not only can they provide new international business skills and networks as well as cheap labour in the services sector, but they may also become an important part of the 'place product'. Immigrant communities that have been accepted by local growth coalitions can be a source of intangible cultural and ethnic capital to a city, facilitating its use of cultural commodities to compete for place advantage in the global cultural economy. In short, immigrant Otherness becomes an essential element of place differentiation. Such Otherness not only provides for attractive visitor experiences, but it may also promote broader 'brand values' of security, harmony, cosmopolitanism and multiculturalism. These, in turn, may attract more general capital flows and certain sought-after immigrant groups to the city.

This does not mean that such commodification processes are ideal – far from it. Yet tourism is the industry of cultural commodification par excellence, and the debates in City Hall on place promotion and marketing and on global competition are dominated by managerialist notions of what constitutes good urban governance, not by criticisms aimed at development and growth coalitions. One has to acknowledge that fact, even though one may wholeheartedly agree with Hewison (1991: 175, italics in original), who argued that:

> the time has come to argue that commerce is *not* culture, whether we define culture as the pursuit of music, literature or the fine arts, or whether we adopt Raymond Williams' definition of culture as 'a whole way of life'. You cannot get a whole way of life into a Tesco's trolley or a V & A Enterprises shopping bag.

The reality is that in terms of urban regional development, the commodification of identity is now an intrinsic component of place promotion. Its validity is largely unquestioned outside the realms of the academy. In a global, fluid world where capital and people are increasingly mobile, 'success' for cities is now often seen in terms of what benefit they can derive from their cosmopolitan and diasporic nature. So rather than rejecting these commodification tendencies outright, students of tourism, migration and place-making need to develop more cogent short-term and long-term policy arguments. These should aim at improving not just the lot of the cities themselves, but also of the people who live there, including the immigrant groups on which this book is focused. Arguably, an important prerequisite for this, at least to the present authors, is to view tourism and migration within the broader scope of global mobility – understanding the reasons why cities and other places seek to attract and retain the geographically mobile, and the means they adopt to achieve this.

Notes

1 'Drop in tourism has a ripple effect in SF', *Asian Week*, 14 December 2001.
2 Oral information from Aart van Boeijen, project developer in this area, to the authors.
3 The literature is inconclusive about the use of concepts such as 'immigrant entrepreneurship' and 'ethnic entrepreneurship' (cf. Rath 2002: 23–4).

References

Abrahamson, M. (1995) *Urban Enclaves: identity and place in America*, New York: St Martin's Press.

Adler, P.A. and Adler, P. (2004) *Paradise Laborers: hotel work in the global economy*, Ithica and London: ILR Press.

Aitken, C. and Hall, C.M. (2000) 'Migrant and foreign skills and their relevance to the tourism industry', *Tourism Geographies*, 2(1): 66–86.

Anderson, K.J. (1988) 'Cultural hegemony and the race-definition process in Chinatown, Vancouver: 1880-1980', *Environment and Planning D: Society and Space*, 6: 127–49.

Anderson, K.J. (1990) 'Chinatown re-oriented: a critical analysis of recent redevelopment schemes in a Melbourne and Sydney enclave', *Australian Geographical Studies*, 28: 131–54.

Anderson, K.J. (1995) *Vancouver's Chinatown: racial discourse in Canada, 1875–1980*, Montreal: McGill–Queens University Press.

Ashworth, G.J. and Voogd H. (1990) *Selling the City: marketing approaches in public sector urban planning*, London: Belhaven Press.

Cartier, C. and Lew, A.A. (eds) (2004) *Seductions of Place*, New York: Routledge.

Castells, M. (1989) *The Informational City: information, technology, economic restructuring and the urban-regional process*, Oxford: Blackwell.

Castells, M. (1996) *The Rise of the Network Society*, Oxford: Blackwell.

Chang, T.C. (1997) 'From "instant Asia" to "multi-faceted jewel": urban imaging strategies and tourism development in Singapore', *Urban Geography* 18(6): 542–62.

Coles, T. and Timothy, D. (eds) (2004) *Tourism and Diaspora*, London: Routledge.

Collins, J. and Castillo, A. (1998) *Cosmopolitan Sydney: explore the world in one city*, Sydney: Pluto Press.

Conforti, J.M. (1996) 'Ghettos as tourism attractions', *Annals of Tourism Research* 23(4), 830–42.

de Gruiter, K. (2000) *West Kruiskade*, Rotterdam: WijkOntwikkelingsMaatschappij.

de Vries, H. (1998) *O, Wies! 't Is hier zo mooi! Reizen in Amerika*, Amsterdam: Contact.

Denker, J. (2003) *The World on a Plate: a tour through the history of America's ethnic cuisinos*, Boulder, CO: Westview Press.

Duncan, J.S. and Duncan, B. (1988) '(Re)reading the landscape', *Environment and Planning D: Society and Space,* 6: 117–26.

Engelen, E. (2001) 'Breaking in and breaking out: a Weberian approach to entrepreneurial opportunities', *Journal of Ethnic and Migration Studies,* 27(2): 203–23.

Fainstein, S.S. and Gladstone, D. (1999) 'Evaluating urban tourism', in D.R. Judd and S.S. Fainstein (eds) *The Tourist City,* New Haven: Yale University Press.

Fainstein, S.S. and Judd, D.R. (1999) 'Global forces, local strategies and urban tourism', in D.R. Judd and S.S. Fainstein (eds) *The Tourist City,* New Haven: Yale University Press.

Fainstein, S.S., Gordon, I. and Harloe M. (eds) (1992) *Divided Cities: New York and London in the contemporary world,* Oxford/Cambridge: Blackwell.

Featherstone, M. (1991) *Consumer Culture and Postmodernism,* London: Sage.

Florida, R. (2002) *The Rise of the Creative Class: and how it's transforming work, leisure, community, & everyday life,* New York: Basic Books.

Frenkel, S. and Walton, J. (2000) 'Bavarian Leavenworth and the symbolic economy of a theme town', *Geographical Review,* 90(4): 559–84.

Gotham, K. (2002) 'Marketing Mardi Gras: commodification, spectacle and the political economy of tourism in New Orleans', *Urban Studies,* 39(10): 1735–56.

Granovetter, M. (1995) 'The economic sociology of firms and entrepreneurs', in A. Portes (ed.) *The Economic Sociology of Immigration: essays on networks, ethnicity, and entrepreneurship,* New York: Russell Sage Foundation.

Hall, C.M. (2005) *Tourism: rethinking the social science of mobility,* Harlow: Prentice Hall.

Hall, C.M. (2007) *Introduction to Tourism,* 5th edn, Melbourne: Hospitality Press.

Hall, C.M. and Williams, A.M. (2002) *Tourism and Migration: new relationships between production and consumption,* Dordrecht: Kluwer Academic.

Halter, M. (2000) *Shopping for Identity: the marketing of ethnicity,* New York: Schocken.

Hewison, R. (1991) 'Commerce and culture', in J. Corner and S. Harvey (eds) *Enterprise and Heritage: crosscurrents of national culture,* London and New York: Routledge.

Hoffman, L.M. (2003a) 'Revalorizing the inner-city: tourism and regulation in Harlem', in L.M. Hoffman, S.S Fainstein and D.R. Judd (eds) *Cities and Visitors: regulating people, markets, and city space,* Oxford: Blackwell.

Hoffman, L.M. (2003b) 'The marketing of diversity in the inner city: tourism and regulation in Harlem', *International Journal of Urban and Regional Research,* 27(2): 286–99.

Hoffman, L.M., Fainstein, S.S. and Judd, D.R. (eds) (2003) *Cities and Visitors: regulating people, markets, and city space,* Oxford: Blackwell.

Hope, C.A. and Klemm M.S. (2001) 'Tourism in difficult areas revisited: the case of Bradford', *Tourism Management,* 22: 629–35.

Jones, T., Barrett, G. and McEvoy, D. (2000) 'Market potential as a decisive influence on the performance of ethnic minority business', in J. Rath (ed.) *Immigrant Businesses: the economic, political and social environment,* Basingstoke/New York: Macmillan/St Martins Press.

Judd, D.R. and Fainstein, S.S. (eds) (1999) *The Tourist City,* New Haven: Yale University Press.

Jules-Rosette, B. (1994) 'Black Paris: touristic simulations', *Annals of Tourism Research,* 21(4): 679–700.

Kearns, G. and Philo, C. (eds) (1993) *Selling Places: the city as cultural capital past and present,* Oxford: Pergamon.

Kloosterman, R. and Rath, J. (2001) 'Immigrant entrepreneurs in advanced economies: mixed embeddedness further explored', *Journal of Ethnic and Migration Studies,* 27(2): 189–202.

Kloosterman, R. and Rath, J. (eds) (2003) *Immigrant Entrepreneurs: venturing abroad in the age of globalization*, Oxford and New York: Berg.

Knecht, M. and Soysal, L. (eds) (2005) *Plausible Vielfalt: wie der Karneval der Kulturen denkt, lernt und Kultur schafft I*, Berlin: Panama Verlag.

Kumcu, A. (2001) *De fil en aiguille: genèse at déclin des ateliers de confections turcs d'Amsterdam*, Amsterdam: Thela Thesis.

Landry, C. (2000) *The Creative City: a toolkit for urban innovators*, London: Earthscan Publications.

Lash, S. and Urry, J. (1994) *Economies of Signs and Space*, London: Sage.

Lew, A.A. and Wong, A. (2002) 'Tourism and the Chinese diaspora', in C.M. Hall and A.M. Williams (eds) *Tourism and Migration: new relationships between consumption and production*, Dordrecht: Kluwer.

Light, I. and Gold, S.J. (2000) *Ethnic Economies*, San Diego: Academic Press.

Lin, J. (1998) *Reconstructing Chinatown: ethnic enclave, global change*, Minneapolis: University of Minnesota Press.

McCrone, D., Morris, A. and Kiely, R. (1995) *Scotland: the brand*, Edinburgh: Edinburgh University Press.

Miller, D., Jackson, P., Thrift, N., Holbrook, B. and Rowlands, M. (1998) *Shopping, Place and Identity*, London and New York: Routledge.

Mitchell, K. (1993) 'Multiculturalism, or the united colors of capitalism', *Antipode*, 25(4): 263–94.

Mollenkopf, J.H. and Castells, M. (1991) *Dual City: restructuring New York*, New York: Russell Sage Foundation.

Morgenroth, L. (2001) *Boston's Neighborhoods: a food lover's walking, eating and shopping guide to ethnic enclaves in and around Boston*, Guilford CT: Globe Pequot.

Morris, M. (1995) 'Life as a tourist object in Australia', in M.-F. Lanfant, J.B. Allcock and E.M. Bruner (eds), *International Tourism: identity and change*, London etc.: Sage.

Mumford, L. (1973) *The City in History: its origins, its transformations and its prospects*, Harmondsworth: Penguin.

Narayan, U. (1995) 'Eating cultures: incorporation, identity and food', *Social Identities*, 1(1): 63–86.

Page, S.J. and Hall, C.M. (2003) *Managing Urban Tourism*, Harlow: Prentice Hall.

Prescott, D. and Wilton, D. (and Dadayli, C. and Dickson, A.) (2000) *Visits to Canada: the role of Canada's immigration populations*, RIIM Working Paper Series #00–16, Vancouver, BC: Vancouver Centre of Excellence Research on Immigration and Integration in the Metropolis.

Ram, M. (1998) 'Enterprise support and ethnic minority firms', *Journal of Ethnic and Migration Studies*, 24(1):143-158.

Ram, M., Abbas, T., Sanghera, B. and Hillin, G. (2000) 'Currying favour with the locals: balti owners and business enclaves', *International Journal of Entrepreneurial Behaviour and Research*, 6(1): 41–55.

Rath, J. (ed.) (2000) *Immigrant Business: the economic, political and social environment*, Basingstoke/New York: Macmillan/St Martin's Press.

Rath, J. (ed.) (2002) *Unravelling the Rag Trade: immigrant entrepreneurship in seven world cities*. Oxford and New York: Berg.

Rath, J. (2005) 'Feeding the festive city: immigrant entrepreneurs and tourist industry', in E. Guild and J. van Selm (eds) *International Migration and Security: opportunities and challenges*, London and New York: Routledge.

Roche, M. (1994) 'Mega-events and urban policy', *Annals of Tourism Research* 21(1): 1–19.

Roche, M. (2000) *Mega-events and Modernity: Olympics and expos in the growth of global culture*, London: Routledge.

Sassen, S. (1991) *The Global City: New York, London, Tokyo*, Princeton, NJ: Princeton University Press.

Scott, A.J. (2006) 'Creative cities: conceptual issues and policy questions', *Journal of Urban Affairs,* 28(1): 1–17.

Selby, M. (2004) *Understanding Urban Tourism: image, culture and experience*, New York: I.B. Tauris.

Shaw, S., Bagwell, S. and Karmowska J. (2004) 'Ethnoscapes as spectacle: reimaging multicultural districts as new destinations for leisure and tourism consumption', *Urban Studies*, 41(10): 1983–2000.

Thrift, N. and Olds, K. (1996) 'Refiguring the economic in economic geography', *Progress in Human Geography* 20(3): 311–37.

Timothy, D. (2002) 'Tourism and the growth of urban ethnic islands', in C.M. Hall and A.M. Williams (eds) *Tourism and Migration: new relationships between consumption and production*, Dordrecht: Kluwer.

Tsu, T.Y. (1999) 'From ethnic ghetto to "Gourmet Republic": the changing image of Kobe's Chinatown in modern Japan', *Japanese Studies*, 19(1): 17–32.

Valle, V.M. and Torres, R.D. (2000) 'Mexican cuisine: food as culture', in V.M. Valle and R.D. Torres (eds), *Latino Metropolis*, Minneapolis/London: University of Minnesota Press.

van Heek, F. (1936) *Chineesche immigranten in Nederland*, Amsterdam: J. Emmering's Uitgevers Mij.

Webster, F. (2001) 'Re-inventing place: Birmingham as an information city?', *City*, 5(1): 27–46.

Wong, K.S. (1995) 'Chinatown: conflicting images, contested terrain', *MELUS*, 20(1): 3–15.

World Travel and Tourism Council (2005) *Travel & Tourism Sowing the Seeds of Growth: the 2005 travel & tourism economic research: executive summary*, London: World Travel & Tourism Council.

Yeung, H. (2002) *Entrepreneurship and the Internationalisation of Asian Firms: an institutional perspective*, Cheltenham: Edward Elgar.

Yeung, H. and Olds, K. (eds) (2000) *The Globalization of Chinese Business Firms*, Basingstoke: Macmillan.

Zhou, M. (1992) *Chinatown: the socioeconomic potential of an urban enclave*, Philadelphia: Temple University Press.

Zukin, S. (1991) *Landscapes of Power: from Detroit to Disney World*, Berkeley: University of California Press.

Zukin, S. (1995) *The Cultures of Cities*, Cambridge MA and Oxford: Blackwell.

Zukin, S. (1998) 'Urban lifestyle: diversity and standardization in spaces of consumption', *Urban Studies*, 35: 825–39.

Part I
Immigrant entrepreneurs

2 Making the new economy

Immigrant entrepreneurs and
emerging transnational
networks of international
education and tourism in Seoul
and Vancouver

Min-Jung Kwak and Daniel Hiebert

International population movements have been a key feature of globalization. With the rapid economic restructuring of recent years, most urban
centres in advanced economies have been transformed by a significant
level of global migration (Castles and Miller 2003). Migrants arrive
through different means and under different administrative categories,
depending on the purposes and conditions of their travel. They include
people working in unskilled jobs, asylum seekers, highly skilled workers,
entrepreneurs, students and tourists. For most immigrant-accepting countries, the four latter groups have been considered more welcome as
sources of labour and innovation – and, of course, consumer spending –
and have therefore been favoured with friendlier migration policies. This
has been the case in Canada too. Since the mid-1980s, the numbers of
'economic-class immigrants', foreign students and tourists have grown
significantly in major Canadian cities.

While these movements are often discussed as concurrent outcomes of
globalization, there has been little effort to examine them as mutually
reinforcing phenomena. Tourism has been categorized as a movement
more or less divorced from other, more 'purposeful' migrations (such as
temporary migration for work or education and permanent migration).
While much research has been done in the separate disciplinary areas of
tourism and migration, there has been little effort to conceptualize the
interrelationship between them (Williams and Hall 2000). In this chapter,
we examine some of the 'messy' dynamics of different migrant flows
between South Korea and Canada. We do this through a case study of
Vancouver, in the province of British Columbia (BC), focusing more
particularly on its Korean-origin community. We thereby situate immigrant entrepreneurs both in the circuits of tourism and in other forms of
temporary and permanent migration, and we discuss their active roles in
connecting what are usually seen as relatively autonomous processes.
This, in turn, leads us to challenge the assumed boundaries between the
production and consumption of cultural tourism. We also examine the

extent and ways in which nation-states regulate (and deregulate) the operational circuits of tourism and migration. Finally, we discuss the unpredictable future of the new economic opportunities emerging in tourism and educational migration.

Conceptual framework

Tourist activities are conceptually understood as an escape from routine in a search for 'other' experiences (Cloke 2000). Tourists are thus seeking transient consumption of aesthetic 'Otherness' in unfamiliar nature and/or culture. Riding on the developments in transportation and communication technologies, the search for Otherness has tremendously expanded its territorial boundaries. As tourism grew into an increasingly important aspect of the global economy, a large volume of research projects on the subject emerged, contributing to our practical understanding of these activities. Until the 1990s, however – before social science was transformed by what has come to be known as the 'cultural turn' – most tourism studies focused on the consumption side of tourism. The patterns, sites, impacts and behaviours associated with tourism have been studied extensively. Meanwhile, the 'production' of these sites has largely been taken for granted and assumed to be a gift of nature or, in urban cases, the result of fortunate historical circumstances.

Drawing on the critical views of cultural studies, recent critics of these foundational studies have called for more nuanced attention to the issues of tourism production. Britton (1991), for example, argued that tourism and travel have been treated as discrete economic subsystems for too long, and that researchers have ignored the links between tourism and structural, economic, social and cultural dimensions of capitalist accumulation (see also Zukin 1990). By studying tourism as a production system that builds and markets 'places and people as a package', researchers have begun to problematize the status of 'authenticity' (Hughes 1995). From this point of view, the commodification of culture in tourist sites and spectacles is part of a social construction of reality and place, and is characterized by racialized representations that are the product of uneven power relations. This logic has also motivated new interpretations of urban cultural landscapes, such as 'Chinatowns' in North American and European cities (e.g. Anderson 1995; Oakes 1993). Key to this work is the idea that sites do not come to be valued by tourists 'naturally', but instead are given meaning through cultural practices.

The 'commodification of culture' is an important concept here. We hope to develop an understanding of tourism that not only takes the production and the consumption of touristic experiences into account, but also the *interaction* between these processes (see also Ateljevic 2000). We will therefore elucidate the role of migrants in circuits of international tourism both as international tourists and as part of the local production of cultural resources. But we emphasize that these developments are mutually constitutive.

Our second analytical theme focuses on the interplay of human agency and structural forces in tourism, and we situate our analysis within a broader context of political economy. This leads us to consider the structural conditions that enable the utilization of cultural capital on different scales, both global and local. Some commentators accept the premise that nation-states have become powerless in the context of irreversible global market forces (Ohmae 1993). We see the opposite. The relationship between the regulatory practices of the state and the development of markets now seems tighter than ever in many ways (Freeman and Ögelman 2000; Kloosterman 2000). In the most general terms, we agree with the view that markets are hardly 'free' at all, and that they instead exist within – and are defined by – a plethora of regulations that govern employment relations and trade systems, that distinguish between legal and illegal products, and so on (Engelen 2003). We focus on more specific matters, of course, and we seek to show that the opportunities embraced by Korean entrepreneurs in Vancouver's tourism and educational sectors are created, or at least supported, to a large degree by activities of the Canadian and South Korean states.

Thirdly, following the lead of critical scholarship (whether arising out of feminism, post-structuralism or anti-racist research), we want to show that taken-for-granted categories frequently conceal important social processes. While we emphasize the role of government in creating a welcome environment for the development of export education, it is equally important to note that many aspects of the emerging system defy the categories – and the regulatory practices – of the state. Ironically, the very regulations that enable this type of entrepreneurship are sometimes circumvented by it. This point has been made by scholars studying transnationalism since the pioneering work of Glick Schiller, Basch and Blanc-Szanton (1992, 1995; see also Vertovec 1999), and it is made most emphatically by Bailey (2001), who argues that traditional dichotomies and categories such as forced versus voluntary migration or temporary versus permanent migration are invalid in the face of the flexible strategies employed by migrants – and, we would add, by states. We examine the ways in which immigrants, students and tourists acquire and utilize cultural capital. We also demonstrate how individuals develop complex trajectories of migration based on their assessment of changing economic opportunities.

Finally, our analysis is guided by an important question that animates scholarship in another field. In the study of immigrant entrepreneurship so far, research has focused mostly on immigrant participation in traditional, low-level, economic activities, such as restaurants, laundries, garment production, taxi operations and the like (Collins *et al.* 1995; Light and Bonacich 1988; Rath 2002; Waldinger and Lichter 2003; some exceptions are Li *et al.* 2002; Lo forthcoming; Wang 1999; Zhou 1998). Instead, we explore immigrants who are developing opportunities at the heart of the knowledge-based economy, in tourism and international education. We can thereby observe the active agency of immigrant enterprise in the 'branding' of Vancouver as a gateway city. We ask whether the recent shift of immigrant enterprise into these sectors represents a real departure from the rather limited opportunities for immigrant entrepreneurship in more traditional fields. More particularly: does the shift from low-level to high-level service provision generate more stable opportunities for immigrant entrepreneurs?

As is increasingly common in contemporary migration studies, we have pursued several forms of information to make our case. We have assembled both census data and administrative data on landings in Canada to build a portrait of the Korean-origin group in Greater Vancouver. Second, we interviewed 17 key informants in the Korean Canadian community. These include three social workers who provide counselling to Korean immigrants and temporary residents in Vancouver, two representatives of Korean Canadian community associations, ten Korean Canadian immigrant entrepreneurs in tourism and international education sectors, one Korean Canadian real estate agent and one investment advisor. All the interviews were carried out in Korean (by Min-Jung Kwak) and translated and transcribed into English for use by both researchers.

Korean immigration and visits to Canada and Vancouver

During the long period of racialized immigration policy in Canada, until the mid-1960s, very few Koreans were admitted. The first period of Korean settlement thus began in the 1960s and involved the entry of skilled workers and their families. Changes in family reunification policy in 1976 led to a lull in immigration from Korea for the next ten years – a time when Koreans migrated instead to the USA in large numbers (see Light and Bonacich 1988). Many Koreans entering Canada during the early two decades quickly turned to entrepreneurial activities, establishing a pattern of economic incorporation still relevant today. The third and still current period began in the mid-1980s, when the Canadian government doubled its immigrant targets and put heavier emphasis on attracting business immigrants (notably by creating a new class of 'investor immigrants'). In recent years (1999–2001), Korea has become the fifth largest source of immigrants to Canada. Since the late 1990s, Canada has become a more important destination for Korean immigrants than the United States.

The growth in Korean immigration to Canada has been more than matched by increases in the numbers of Koreans coming to Canada as tourists, visitors and temporary residents. In 2002, according to the agency Tourism British Columbia, South Korea accounted for the fourth largest number of individuals cleared through customs in British Columbia (after

USA, Japan and UK). This is a remarkable statistic, considering the high degree of transnationalism associated with the overseas Chinese (e.g. Ong 1999; Yeung and Olds 2000), and the fact that the Vancouver Korean population is far smaller than its Chinese- and Indian-origin counterparts. Altogether about 150,000 Koreans came to Canada as tourists in 2002, and 100,000 cleared customs in BC (Tourism British Columbia 2002).

In addition to tourists, the number of Koreans on temporary visas in Canada, and especially in Vancouver, has been steadily increasing too. These include special categories of workers, such as employees of multinational corporations, a few other minor categories, and students. The rapid growth in this latter category began in the 1990s, after Koreans were granted a visa exemption in 1994.[1] There was a pronounced dip in student entries to Canada in 1998, coinciding with the 'Asian flu', but the number resumed its upward momentum the following year. Since then (the late 1990s), Korea has become the largest source of international students in Canada; over 100,000 temporary student visas were granted to Koreans between 1998 and 2003.[2] However, we have also learned that perhaps half of the Koreans studying in Canada do not bother to obtain a student visa, but simply enter Canada as tourists and either complete their studies within six months or renew their tourist visas if they are engaged in longer programmes of study (see also VEDC 2003). Thus, there are probably at least 30,000 Korean students arriving in Canada each year. About 40 per cent are destined for Vancouver (CIC 2003).

The unprecedented flows of tourists and migrants from Korea have had significant impacts on the resident Korean Canadian community. A number of Korean residential and commercial concentrations have emerged, adding new culturally distinct precincts to an already diversified urban landscape.

The commodification of culture: beyond multiple binaries

The advance of global capitalism has accelerated the commodification of nearly every form of good or value (Nash 2000). In the circuits of contemporary tourism, cultural resources have been imbued with economic potential. From high culture to popular street fashion, cultural resources are developed and put on the market for sale just like other, more ordinary consumer goods. Tourist ventures that offer consumers the products and experience of exotic splendour have become easier to market, because time and space constraints are partly overcome by cheaper travel (in relative terms) and communication. The scope for the commodification of cultural products and values has thus become effectively global, generating new sources of international investment and production that cross borders (Zukin 1990). There is more tourism from advanced economies to the developing world, but the emerging system also includes new developments in advanced economies. As migration to post-industrial centres has accelerated, cultural diversity within them has also become a source of commodification. This can be seen in myriad ways, from Mexican food vendors in suburban shopping malls to martial arts centres in downtown 'Chinatown' precincts. In these contexts, ethnic entrepreneurs are actively cashing in on their knowledge and skills of relevant cultures.

The ways in which cultural resources are bought and sold are worthy of further consideration. Bourdieu ([1979] 1984) argued that consumption preferences are closely related to the process of social reproduction. Consumption is also closely tied to identity. In the process of learning and practising culture, individuals consume cultural resources and accumulate symbolic capital, and in so doing they help to reproduce larger cultural patterns. While Bourdieu explains the process of individual cultural consumption in close relation to hegemonic power, Zukin (1990) and Ong (1999) have modified the theory, challenging this rigid interpretation of cultural capital accumulation. In certain contexts, they argue, cultural capital becomes 'elastic' and 'flexible' as resources are negotiated between and reproduced by different actors. In treatments of cultural capital inspired by Bourdieu, immigrants and ethnic minorities have often been depicted as mere bearers of inherited culture and traditional values. But following Zukin and Ong, we argue that immigrants, tourists and international students can be viewed as flexible cultural capitalists, who actively seek and learn new, 'foreign' cultural experiences and utilize them with a mix of their own. Rather than assuming that their experiences are locked in tradition, or homogenous, we therefore focus on the different constructions of cultural capital utilized by different actors.

Consider first the following personal background information on the settlement process of three of our respondents:[3]

> *Ms Kim* came to Canada in 1991 as an ESL (English as a second language) student. After attending a university-affiliated ESL programme in Edmonton for about six months, she moved to Vancouver and worked as a caregiver in a Canadian family. She recalled that period as arduous but said it was also a good opportunity to learn English and get to know Canadian culture. After four years of caregiver work, she earned permanent residence in Canada and began her own international education agency. Aware of the difficulties she herself experienced as a foreign student in Canada, Ms Kim now offers both technical and emotional support to students. She operates one of the most successful international education agencies in Vancouver today.
>
> *Mr Kang* first visited Canada as a tourist in 1996. Seeking business opportunities and a chance to settle in Canada, he successfully applied as a business immigrant. His wife and children arrived later. After learning how to operate a tourist agency while employed in a friend's firm, he opened his own agency in a suburban Korean commercial district. He now has another branch downtown, which shares office

space with an international education agency – giving his company good exposure to international students. It is a smaller branch managed mainly by his wife.

Mr Sohn arrived in Canada as a business immigrant in 1995. He spent several months attending an ESL school and eagerly sought out good business opportunities. While Mr Sohn was learning English, he soon became aware of inefficiencies in the programme. Knowing that East Asian students, the largest customer base for Vancouver's international education sector, usually have trouble understanding spoken English and are reticent to speak, he decided to open an ESL school designed to offer more specialized and intensive English classes. (The idea was not really unique for Korean students, because many ESL institutes in Korea already operated in the same way.) Since the first class offered in 1996, the school has grown apace. In 2002, it was even visited by Denis Coderre, the Canadian Minister of Citizenship and Immigration, who cited the school as exemplifying a successful venture by business immigrants.

Although Ms Kim, Mr Kang and Mr Sohn arrived in Canada under three different visa categories, they went through a similar process of learning about Canadian economic systems and culture. Their journey did not end as consumers or employees; they went on to become entrepreneurs who actively mediate between Korean and Canadian cultures. Furthermore, in their businesses they also attract potential tourists and students to Vancouver. They participate in the commodification of culture, pursuing mix-and-match strategies that combine and reproduce cultural resources. The process has not been particularly linear; it is situational and ongoing.

The ways in which Korean entrepreneurs operate their businesses transcend many of the boundaries between categories used within the immigration policy/programme system, and by academics as well. Even the term 'international education industry' implies a far more coherent and contained set of economic processes than is actually the case. In our interviews, many participants distinguished small firms from large ones, explaining that small operations try to recruit clients through highly personalized services offered in a full package of options – including school placement, visa processing, travel arrangements (both for the journey from Korea and for excursions in Canada), arranging home care for students and assisting individuals who wish to immigrate to Canada. They contrasted this flexible, personal attention with the more formal business practices of larger firms. However, when speaking with the owners and managers of the larger firms, especially those that are linked in multi-node business networks, we found that they offer approximately the same range of services, though often at a higher price. Hence, both small- and middle-sized firms offer such a wide range of services to clients that they are difficult to classify. Are they education institutions, travel agencies, real estate agents or immigration consultants? In many cases, they are all of these things.

Another form of blurred boundaries has already been hinted at in the previous paragraph and in the three biographical notes presented earlier: who exactly are the clients of these institutions? According to an owner of one travel agency:

> Our main customers are Korean immigrants living here, plus some visa students and their families. Most of our package tour customers are new immigrants or their relatives from Korea. Often, many goose parents[4] visit their children here, and go travelling around during the summer or in holidays. In that case, they use our package tour as well. That is common. Many parents eventually apply for permanent immigration too. We don't do immigration business but I've noticed that about half the families with children here eventually decide to immigrate to Canada themselves.[5]
>
> (Ms Yoon, travel agency owner)

The above quotation highlights the links we want to reveal in this paper: immigrants attract visitors, and visits reinforce the desire to immigrate. The immigration and tourism movements, together with the desire of Koreans to obtain foreign educational credentials and to pursue cosmopolitan experiences as a means of gaining cultural capital, provide a base for an emergent form of international tourism which includes globalized education. Students and tourists augment the market opportunities of a growing Korean Canadian network of small firms in Vancouver. These interlocking markets are further illustrated in the following quotes:

> *MJK*: As you said, students travel about while they are in Canada to study. Do you have any connections with travel agencies?

> We have our own travel agency. Right now it's located in one of our branches downtown. But I'm planning to move it to its own office and give it an education function. If we deposit C$5000 [about €3000], we can open an office. So it's not something really difficult to do. Students come to Canada to study, but they usually like to travel too. And I think that's important for them. Our travel agency will help them explore Canada's different sights. Later they'll come back on their honeymoon or with their family.
>
> (Ms Kim, international education agency owner)

> I know that Canada considers international students to be potential immigrants for the near future. I think that makes sense. To prepare for this business opportunity, I called over a friend in Korea to help

me here. I hired him in our agency and am helping him get through the University of British Columbia's immigration consultancy programme. That way our agency will have a fully trained immigration consultant. This will help many foreign students who wish to stay in Canada.

(Mr Chung, international education agency owner)

My business focuses on selling tour packages to Korean tourists in the US and Korea. In the US, I have more than one partner agency in every major city. For the Korean market, I largely depend on my own social networks – families, relatives and, you know, friends. They also send their children to be educated here. I take care of those cases too. During vacation season, those children's families also visit Canada. They are quite an important income source for my business.

(Mr Kang, travel agency owner)

Recall the quotation from Ms Kim: 'Students come to Canada to study, but they usually like to travel too'. The categories of students, tourists and immigrants are often conflated. For Ms Kim, her clients are tourists as much as they are students. As the three biographical notes tell us, students and tourists may well become immigrants too. Three of the ten entrepreneurs we interviewed originally came to Canada as students and one as a tourist. Two former students have become permanent residents and one is currently undergoing the application process. Min-Jung Kwak, one of the authors of this chapter, set in on two immigration information seminars held in Vancouver – one offered jointly by Citizenship and Immigration Canada and the South Korean consulate and one by a privately owned consultancy agency – and observed a mutual interest by students, their parents and both states with regard to permanent resident status. Other researchers in Canada (Ries and Head 1998) and New Zealand (Kang and Page 2000) have similarly pointed out mutually reinforcing relationships between immigration and tourism, each leading to more of the other. The four owners/managers of travel agencies interviewed for this project all corroborate this point.

This last form of 'blurring' is the most difficult to assess, and was not part of the conversation with many of our interviewees. A significant proportion of the money that changes hands in the tourism and related sectors does so without documentation and is therefore untaxed. This is particularly true of the home-stay part of the business, which amounts to tens of millions of dollars each year: many homeowners accept room and board fees without issuing receipts and without declaring the income generated from this service to the relevant authorities. The 'paper trail' on these transactions is minimal, as the money is paid by tourists and international students who withdraw it from banks located in other countries. It is interesting to note that this practice is common knowledge. The income deriving from home-stay fees was even included in a recent official attempt to estimate the economic impact of export education for the Vancouver economy (VEDC 2003), even though much of it is hidden. Thus, even though the bulk of the export education industry is a fully registered part of the formal economy, some zones of it exist more informally.

Political economy of cultural tourism: markets versus (de)regulations

Since the 1960s, the tourism industry has become a major sector of the Canadian economy. For British Columbia, the trend has been even more significant than for other parts of Canada. Today, the tourism gross domestic product of BC accounts for nearly one fifth of the national tourism economy.[6] Among many other contributors, South Korea has become a major source of tourists to Canada, and more specifically to BC, in recent years. In 2002, more than 150,000 Koreans visited Canada, and two-thirds of the inflow came through BC customs.[7] In 2001, total spending by Korean tourists was estimated at C$264 million, ranking sixth among the overseas tourist markets (Tourism British Columbia 2001).

International students who come to Canada to learn English or to pursue other educational goals can be considered cultural tourists (Kennett 2002). In general, they enjoy many leisure-related activities, but they also make a systematic investment in cultural capital by consuming Canadian education and culture. Although it is hard to calculate, the economic benefit of the international education industry is significant. A recent study published by the Vancouver Economic Development Commission (VEDC 2003) focused on just one part of the export education sector in Greater Vancouver, English language training. Estimates extrapolated from these figures suggest that international students spend C$500 million (around €300 million) in Vancouver on tuition fees and housing, and an additional C$260 million on entertainment, transportation and retail purchases. Moreover, the ESL schools employ some 2000 workers in Greater Vancouver. These are impressive numbers, though it is hard to know how accurate they are. Regarding the scale of the industry, one of our respondents explains:

> Currently we have a total of 22 branch offices. Korea has the most branches, eight in all. Of the others, six are in Vancouver and two are in other Canadian cities, Toronto and Calgary. We also have one in Mexico and two in Japan. Basically, our international branches recruit students and send them here [to study in Vancouver]. Locally we provide a variety of services that students need here.
>
> (Ms Kim, international education agency owner)

In this case, a Korean company has arranged its practices globally, with the ultimate goal of recruiting students to Vancouver from several parts of the world, and sharing profits between Korean and Canadian operations. Another entrepreneur, who immigrated to Canada in the early 1990s, confirmed this as a popular strategy:

> Marketing is important. We do traditional marketing [through education agencies] as well as our own. Although Korean students are our main customers, we also have our own marketing offices in Japan, China and Brazil, in addition to Korea. We also maintain business partnerships with 300 agencies in Korea.
>
> (Ms Choi, ESL school coordinator)

The final point in the above quotation gives an indication of the scale of the South Korean overseas education system. In 2001, there were nearly 7000 tourist agencies operating in South Korea, and some 40 per cent of them were specialized in outbound travel (Kim 2001). About 800 of the latter provided services linked to international education. The mixed service provision of firms in Canada is hence also apparent in Korea. The number of firms suggests a high level of economic fragmentation in terms of ownership, as well as the existence of network strategies.

● ● ●

The rapid development of tourism markets is not only driven by global market forces, but it is also redefined by the regulatory practices of states. The growth of the tourism industry as a whole, and the international education industry in particular, in both South Korea and Canada should therefore be viewed as an outcome of a complex interplay between various market forces and multiple governmental bodies.

On the South Korean side, the national government's effort to develop and regulate the tourism industry has been notably systematic. Responding to the Korean International Tourism Organization Law, the state founded the Korean National Tourism Organization (KNTO) in 1962. It is a government organization that promotes inbound tourism, develops resources and trains tourism-related personnel (KNTO 2005). Including its recently established Toronto office, the KNTO operates 20 overseas offices in 13 countries around the world to attract more foreign tourists to South Korea. Within the state administration, the Ministry of Culture and Tourism (MCT) in Seoul has been responsible for introducing tourism-related policies in consultation with the KNTO. Just as South Korea's economic growth has been largely driven by its strong government, the tourism industry has been protected and nurtured by the state.

With the 1989 foreign tourism liberalization policy, major governmental restrictions on overseas travel were lifted, and the South Korean tourism industry entered a new phase (Im and Song 1992). In actual fact, the South Korean government's stance on tourism, and on international education in particular, has become rather ambiguous since the liberalization policy was introduced. The number of outbound tourists and international students from South Korea has been increasing significantly, causing a serious trade deficit. The outflow of students has attracted the most attention, because it implies much higher costs and other social impacts. Note that data on education abroad are gathered by the National Institute for International Education Development, affiliated with the Ministry of Education and Human Resources Development. It keeps track of foreign students in South Korea as well as Korean students abroad and maintains statistics on the annual numbers of Korean students around the world. In collaboration with marketing agencies in host countries, it also provides Koreans with official information about studying abroad.

For some time now, the Ministry of Education has officially prohibited K-12 (kindergarten-to-grade-12) pupils from leaving Korea to pursue an education abroad.[8] Yet this rule is routinely broken, and more than 80 per cent of the total cases of international education registered in official data between 1995 and 2001 were unauthorized (MEHRD 2002a). This amounted to 25,000 primary and secondary school pupils. How could so many pupils study abroad when that is officially prohibited? Regulations are weak. Penalties to parents for sending children abroad are insignificant, and there are no laws against Korean companies facilitating this practice. The 500 or so private education agencies in the country do not face sanctions if they provide services to send K-12 pupils abroad in conjunction with their main business of foreign post-secondary education.

The growing demand for K-12 education abroad is worrying to the South Korean government, as it could be interpreted as a widespread lack of confidence in the national education system. Moreover, the fact that some parents can afford to send their children overseas while others cannot is arguably deepening the social polarization of the country and further enhancing the ability of those with middle to high incomes to pass on their privileges to the next generation. This has led to protest from lower-income families. Nevertheless, regulatory activity is unlikely to be stepped up in the near future. One of our interviewees made this point quite forcefully:

> For such a long time, those high-class officials have been sending their kids to the US and Europe. Only in very recent years has it become very popular among ordinary people. If [the authorities] were to regulate international education, that would be wrong. If the Korean government decides to criminalize parents that send their children abroad, that will hurt the authorities and the families the most.
>
> (Mr Chung)

Turning to the Canadian situation, we begin again on the Western side of the Pacific in South Korea, where the Canadian government actively encourages a number of developments related to tourism and export education. Increasingly, English fluency is recognized as essential to be a competitive global citizen. In 2001, more than half of the 150,000 Korean post-secondary students studying outside Korea were located in the USA and Canada (MEHRD 2002b). To respond to the demand, the Canadian Tourism Commission (a crown corporation of the federal government) has opened an overseas office in South Korea, which promotes tourism in advertising campaigns and a Korean-language website. Citizenship and Immigration Canada has opened a Canadian Education Centre in Seoul and takes part in annual education conventions. The Canadian embassy in Seoul encourages trans-Pacific investment, providing advice to Canadian entre-preneurs operating in Korea as well as to Koreans interested in Canada. Part of that interest, of course, is related to Canada's three programmes designed to recruit business immigrants.

However, while these proactive programmes have likely been instru-mental in attracting Korean investment, we believe they are surpassed in significance by a more passive change in policy, the relaxation of the visa requirements for South Korean visitors to Canada in the 1990s. This point is emphasized by Ms Yoon, a Korean who immigrated to Toronto in 1991 but moved to Vancouver to three years later to tap into the possibilities opened by the new regulations:

> It was 1994 and the visa exemption had come into effect between Korea and Canada. We thought the travel business would be good, that's why we opened this business. Well, since 1994, people have indeed started to come. Korean students also began arriving, a real influx. It was much more difficult to get to Canada before, because of the complicated visa procedure.
>
> (Ms Yoon)

As the owner of an education agency in Vancouver put it:

> In my opinion, there is little connection between the size of the Korean community [in Vancouver] and the influx of Korean students. There could be a little, but what has been more important were the Canadian immigration policies that welcomed both immigrants and students. Also, the rapidly developing Korean economy has been an important factor. If the Korean economy hadn't been doing well, the Canadian government would have never opened its doors to Korean people like it did. So I think the most important factor is economic power.
>
> (Ms Kim)

Still on the theme of passive policies, the relative lack of regulation governing educational services, particularly in export education, is also a crucial factor in the development of this sector. The same informant illustrates this point in describing the origins of her education agency:

> I started my business at home, in my small downtown apartment. When I first registered it in 1996, [the BC government] didn't even know what an international education information centre was all about. My agency was one of the first for Korean students. Seeing that my business was successful, many followed my business strategy. There are many agencies now. Some agencies have been set up by my previous employees. I think this is a positive thing. You know, competition usually results in a higher quality of service provision to customers.
>
> (Ms Kim)

A coordinator of ESL programmes in Greater Vancouver adds an important detail:

> Our school has dealt with more than 200 agencies in Vancouver as well as worldwide. There are more than 100 Korean-owned and -managed local agencies in Vancouver and the Lower Mainland that we have business relationships with. These include all the entrepreneurs who run their businesses at home. Most small home-based agencies don't last long. But I can't neglect them. You see, some students prefer to receive more personalized care than systematic service.... Of course there's a difference in commission for small versus large agencies. But we also pay a commission for recruiting a small number of students.
>
> (Ms Song, ESL coordinator)

Ms Song alludes here to an increasingly common practice: local primary and secondary schools that are part of the Canadian public education system pay commissions to private businesses that recruit international students on their behalf. The schools engage both with large, established recruitment firms as well as with tiny ones operating out of the homes of entrepreneurs. The lack of regulations governing the scale of agencies, and the apparent lack of preference for dealing with small versus large agencies, provides opportunities for small firms keen to enter the field of export education. Start-up costs are very low, and regulatory barriers to this type of business are virtually nonexistent.

Where do the 15,000 or so Korean students who are studying in Vancouver at any given point in time find housing? What about short-term visitors and tourists? For the most part, educational institutions arrange with local families to provide room and board for visiting students, and an elaborate 'home-stay' system has developed. Students typically pay C$700–800 (€400–500) per month for room and board if they are living with a non-Korean family, or between C$1000 and C$1500 with a Korean family, where they can expect familiar food and more cultural understanding. It is easy to find both short-term and long-term home-stay advertisements in any ethnic media written in Korean. A real estate agent who settled in Canada in 1975 explained the impact of this system on the housing market:

> Home stay? Oh yes. That's very popular for everyone in Vancouver. Many [established] Koreans look for housing with a lot of extra rooms. In fact, many Indo-Canadian home builders seem to be well aware of this trend. They build appropriate housing for home stay. And probably that is the least risky business of all, I would say, [for Koreans who want to generate self-employment income].
>
> (Mr Cho, realtor)

Students tend to choose non-Korean families when arranging room and board, both to reduce costs and to enhance their language training by inter-acting 'at home' in English. After a few months in that situation, though, many post-secondary students find independent housing, usually with Korean roommates. For this they turn to the local rental market, providing a new source of revenue for landlords. Whatever the situation – home stay or private rental – there are virtually no regulations governing this aspect of the export education system.

One regulation applying to K-12 pupils creates additional scope for immigrant entrepreneurship. Nearly all of the municipal school boards in Greater Vancouver require international pupils to have a legal guardian, for whom their parents typically pay C$4000–6000 per year.[9] Guardians are interlocutors between the pupils, their parents in Korea and the local school system, and they therefore need to communicate in both English and Korean. While the school system requires guardians to have perma-nent residence status in Canada and adequate language proficiency, there are usually no other qualifications or strict responsibilities connected with the position. This relatively permissive regulatory environment has enabled a great deal of economic activity, ranging in scale from small, home-based businesses, owned and operated by a single person, to medium-sized firms.

Immigrant entrepreneurship in the knowledge economy

It is widely known that many businesses started by immigrants and ethnic minority people lead a precarious existence, with high rates of failure, low profits and low incomes relative to investment (see e.g. Li 2000). Several theories have been proposed to account for this lack of success: customers may discriminate against such firms or may patronize them only if they offer rock-bottom prices (Walton-Roberts and Hiebert 1997); immigrants and ethnic minorities may be propelled into self-employment after facing an unwelcoming labour market, even if they are not ready to take up the challenges of entrepreneurship (Teixeira 1998); the firms are in highly competitive sectors of the economy, such as catering, construction or taxi transport (Hiebert 2002; Langlois and Razin 1989; Li 1994; Phizacklea and Ram 1996); or they rely on local markets, but are frequently situated in poor areas whose residents have little purchasing power (Phizacklea and Ram 1996).

We might expect a different set of outcomes for the firms investigated in this chapter. In the first place, we are dealing with education and tourism, hardly a sector traditionally associated with immigrant or ethnic minority enterprise. Secondly, we are dealing with firms which, either by themselves or through networks, reach across international borders and are not overly reliant on a local market. The international education and tourist agencies could, in fact, be considered an emerging form of the transnational immigrant enterprise highlighted in a recent paper by Portes *et al.* (2002). These authors identify four types of transnational immigrant entrepreneurship, none of which quite fits the sectors of international education and tourism explored by us. They are: circuit firms that transfer goods and remittances; enterprises that transfer cultural products between countries; small retail firms catering to the specific needs of migrant communities; and return migrant enterprises, set up upon return to the source country after a sojourn abroad. We would add a fifth type: immigrant-owned firms that facilitate the transnational circulation of people. In any case, the Portes group have found that transnational immigrant entrepreneurs are relatively more successful than those operating traditional immigrant enterprises. Again, this suggests we are likely to encounter a relative degree of success in the international education- and tourism-related businesses run by immigrants.

One theme that emerged in our interviews certainly confirms this expectation. The owners and managers of educational institutions, travel agencies and recruitment firms frequently voiced a degree of pride in their business, which they see as more prestigious than the enterprises typically owned by Korean Canadians, such as corner shops and grocery stores. As Ms Choi put it, 'I'm proud of what I'm doing. Helping students is a spiritually fulfilling profession'. A travel agent also spoke about the resilience of demand for his firm, which seems to exist regardless of economic events:

> We are not much affected by the Korean economy. There's a bit of impact from the bad local economy. But overall our business is very stable. Think about it: Koreans here always visit Korea every year or two. That's their vacation. They don't go to different places, but to Korea instead. Also, Korea is not a place you can't afford to go to because of economic difficulties. Since we have many Korean Canadian customers that regularly visit Korea, we're less likely to be affected by the regional economy.
> (Mr Noh, travel agent)

The rapid growth of the industry is another sign of success potential. As the Vancouver Economic Development Commission reports, there are now nearly 200 businesses in Greater Vancouver dedicated to ESL programmes alone, and these represent just one facet of the industry.

Contrasting with these upbeat signals are other indications that export education might not be a robust sector for entrepreneurship in the long run. We have already seen one such sign – the sharp decline in students triggered by the Asian recession of the late 1990s. This unforeseen and distant economic event had painful, though temporary, implications for Vancouver's export education sector. One of our respondents articulated this worry:

> This business is very vulnerable to many factors. It's sensitive to the world economy, and of course to the Korean economic situation. This year hasn't been too good because of the slow economic situation in Korea. Also, here there was SARS and the Iraq war. People are sensitive to exchange rates, too. So this is a difficult business to maintain consistency in. To be successful, I have to develop new strategies all the time.
>
> (Ms Choi)

The fact that these firms are targeted to the ethnic Korean market presents both advantages and problems, which are summarized by Ms Yoon, a travel agent:

> Korean people have been coming at a steady pace so far. Umm, since we're located right downtown ... we first served Korean consulate people. Our previous office was in the same building where the consulate was located. So consulate people and Canadians were our main customers. But more and more Korean people are coming to Vancouver, and they now make up the major proportion of our business. They generate so much business that we don't even need Canadian staff. Other reasons are that ... the marketing for Canadian travellers is more difficult and the business with them leaves us very little profit. Because there is enough demand from Korean customers, both in ticket sales and package tour sales, we began moving away from the Canadian market. For the last ten years, our sales to Korean customers have kept growing. So I think the growth of the Korean immigrant community in Vancouver has been helpful to my business.
>
> (Ms Yoon, travel agent)

Another travel agent spoke about the difficulty of gaining access to customers from different ethnic backgrounds.

MJK: What about the Vietnamese community? I notice a lot of Vietnamese commercial signs in this area [Kingsway].

Mr Noh: Yes, there's a large Vietnamese population but it's very hard to target the community.

MJK: What about the Chinese?

Mr Noh: We have a Chinese agent working right over there [pointing]. Amy has worked with us for seven years now. She's from Hong Kong. So as you can see, without someone speaking the right language, it's very difficult to target other ethnic communities.

Many Korean international education agencies and ESL schools have tried with some success to recruit customers from different countries and ethnicities. Typically these are larger-scale institutions that are able to hire appropriate staff. Most Korean-owned firms, however, still depend on the Korean market, meaning that their fortunes are linked to the economic situation in South Korea.

Considerable concern was also expressed about rising competition in both the tourist agency and the export education sectors. We can see a familiar scenario operating in both. A few innovative entrepreneurs discover market potential and set up businesses. In so doing, they actually shape the market by offering services where none or few existed before. Their success inspires others to enter the market. Eventually the market becomes crowded with competitors, prices fall, and profit margins are reduced. A few businesses develop successful strategies in these stressful circumstances, but many do not. The lack of regulations, which permits easy entry into an industry, also allows heightened competition between firms that engage in very different business practices:

> Well, I hope government will support the small travel industry more in general. But I also think there should be stricter regulations on qualification. I've seen so many international consultancy agencies enter the travel business now. They're not permitted to sell air tickets, and they're not even qualified to give professional advice to customers in the first place. We [qualified travel agencies] are damaged by excessive competition with unqualified small travel agencies. We also have the big companies to compete with. It's difficult. As a qualified travel agency, you're required to make a deposit of C$20,000 to the travel association. Many unqualified agencies do business without that deposit. So when they go bankrupt, the customers are not compensated. I think government should intervene more firmly.
>
> (Ms Yoon)

Mr Oh, who manages an ESL school, speaks about the situation of recruitment agencies in similar terms:

> However, as competition has gotten worse, agencies began to ask more and more from schools. The result is a price war in this industry.

I think price has to be the last option, not the best option for business management.

(Mr Oh)

A real estate agent put the issue into perspective. He senses that the long-run situation in the 'new' Korean Canadian economy might be very bleak:

I usually advise new immigrants to look at the most popular businesses among Koreans. The typical Korean businesses such as grocery and coffee shops, fast food restaurants and laundry shops are safe choices for them. I always tell them if they can make three to four grand [€1800–2400] a month, they should be happy. I've seen many new immigrants with high entrepreneurial spirit and enthusiasm looking for other business opportunities. They try everything. But I've also seen enough failures. New areas require enormous time and energy. In reality, many new immigrants can't afford to invest that much. Hard work and modest incomes are things you have to put up with in Canada.

(Mr Cho)

Final thoughts

This initial and partial investigation has only touched on the basic story of Korean transnational entrepreneurship in the international education and tourist agency sectors. There are far more issues involved in this process than we have had the scope to discuss, such as the relationship between the actual entrepreneurship in these sectors and the rules and expectations of Canada's business immigration programme. It is clear, though, that we are seeing the outlines of a form of globalization that has been little noticed as of yet, even within the tourism sector. The ways in which tourists, students and immigrant entrepreneurs consume and utilize cultural resources reveal some of the complex dynamics of the accumulation of cultural capital. Many of the activities described here are invisible to the state, and hence go unrecorded in all of the standard sources of data used in Canadian immigration research. True, statistics are kept on student and tourist visas, but even these seemingly straightforward categories are confounded in reality, since students do not necessarily apply for visas.

Ironically, even though some activities are invisible to the state, the state is still the key institution that generates opportunities for export education. Or more accurately we should talk about 'states', as this form of economic development arises partly in the intersecting policies and programmes of the South Korean and the Canadian governments. In this field, one cannot speak about markets without speaking about politics, and vice versa. It is also clear that the entrepreneurship we have explored is highly flexible and spans several sectors of the economy that are usually regarded as quite separate: tourism, education and housing. The linkages between these activities are being generated by immigrants, who are assertively creating a new branch of the 'knowledge economy'.

There are, however, indications that the industry is economically precarious. We have seen that a downturn in the South Korean economy in 1998 led to a precipitous drop in the number of student visas issued and to the closure of some private schools. Severe competition and the difficulties of market development beyond a co-ethnic clientele are a familiar story in typical immigrant enterprises. In other words, the economic opportunities we have described in this paper are unstable, and they could either continue or vanish through factors beyond the control of the entrepreneurs who have established and run the industry. In the future we could conceivably see export education and its tourism linkages as just another marginalized sector of the economy associated with immigrant enterprise and immigrant workers. For the moment, though, the emerging transnational education sector is contributing significantly to the well-being of the Korean immigrant community in Vancouver, and indirectly to the rest of the population as well.

Notes

1 Since the 1994 modifications, South Koreans can visit Canada for up to six months without obtaining a special visa (that is, they simply receive a stamp in their passport on arrival to Canada).
2 Administrative data from Citizenship and Immigration Canada (CIC/FOSS, data warehouse cubes); we thank CIC Pacific Region for providing this figure.
3 At interviewees' request, pseudonyms were used to protect their identity. The background information on Mr Sohn was obtained from Mr Oh, a Korean coordinator of the ESL school.
4 Koreans use the term 'goose parent' in much the same way that Chinese people speak about 'astronaut families'. Goose parents are frequent fliers who shuttle between South Korea, where they work, and Canada, where their children are.
5 Aitken and Hall (2000) argue that the skills of foreign-born workers are vital to the success of the tourism industry. Clearly, Ms Yoon's business is based on her ability to interact with Korean customers.
6 'Tourism industry and culture: what are the problems?' [in Korean], *Coreamedia* (Vancouver), 12 July 2003.
7 This statistic includes neither immigrants nor temporary visa holders (e.g. international students and temporary workers).
8 This rule was relaxed in the latest revision of international education policy in 1998. Adolescents are now allowed to obtain a secondary school education elsewhere, provided they are evaluated as outstanding and recommended by their school head; it is still unlawful for primary and middle-school pupils to do so. The Ministry of Education is also stepping up its efforts to recruit foreign students to come to Korea for post-secondary education.
9 Officially, pupils in primary school must be accompanied by at least one parent, but this rule is often ignored in practice. The amount paid for guardian services varies, because some school boards have more elaborate requirements than others. The Surrey school board, for instance, mandates that guardians must be at least 25 years of age and either landed immigrants or Canadian citizens. That is not the case under other school boards.

References

Aitken, C. and Hall, C.M. (2000) 'Migrant and foreign skills and their relevance to the tourism industry', *Tourism Geographies*, 2(1): 66–86.

Anderson, K.J. (1995) *Vancouver's Chinatown: racial discourse in Canada, 1875–1980*, Montreal: McGill–Queens University Press.

Ateljevic, I. (2000) 'Circuits of tourism: stepping beyond the "production/consumption" dichotomy', *Tourism Geographies*, 2(4): 369–88.

Bailey, A. (2001) 'Turning transnational: notes on the theorization of international migration', *International Journal of Population Geography*, 7: 413–28.

Bourdieu, P. [1979] (1984) *Distinction: a social critique of the judgement of taste*, trans. R. Nice, Cambridge MA: Harvard University Press; originally published as *La Distinction: critique sociale du jugement*, Paris: Editions de Minuit.

Britton, S. (1991) 'Tourism, capital and place: towards a critical geography of tourism', *Environment and Planning D: Society and Space*, 9(4): 451–78.

Castles, S. and Miller, M.J. (2003) *The Age of Migration: international population movements in the modern world*, 3rd edn, New York: Guilford Press.

CIC (Citizenship and Immigration Canada) (2003) 'Foreign students in Canada', available HTTP: <www.cic.gc.ca/english/pub/index-2.html#statistics> (accessed 1 March 2005).

Cloke, P. (2000) 'Tourism', in R.J. Johnston, D. Gregory, G. Pratt and M.J. Watts (eds) *The Dictionary of Human Geography*, Oxford: Blackwell.

Collins, J., Gibson, K., Alcorso, C., Tait, D. and Castles, S. (1995) *A Shop Full of Dreams: ethnic small business in Australia*, Sydney and London: Pluto Press.

Engelen, E. (2003) 'How to combine openness and protection? Citizenship, migration and welfare', *Politics and Society*, 31(4): 503–36.

Freeman, G. and Ögelman, N. (2000) 'State regulatory regimes and immigrants' informal economic activity', in J. Rath (ed.) *Immigrant Business: the economic, political and social environment*. Basingstoke/New York: Macmillan/St Martin's Press.

Glick Schiller, N., Basch, L. and Blanc-Szanton, C. (1992) 'Towards a transnational perspective on migration: race, class, ethnicity and nationalism reconsidered', *Annals of the New York Academy of Sciences*, 645: 1–24.

Glick Schiller, N., Basch, L. and Szanton Blanc, C. (1995) 'From immigrant to transmigrant: theorizing transnational migration', *Anthropology Quarterly*, 68: 48–63.

Hiebert, D. (2002) 'Economic associations of immigrant self-employment in Canada', *International Journal of Entrepreneurial Behaviour and Research*, 8: 127–40.

Hughes, G. (1995) 'Authenticity in tourism', *Annals of Tourism Research*, 22(4): 781–803.

Im, E. and Song, S. (1992) 'A study on the effect of the liberalization of overseas travel on outbound travel: Korea and Japan' [in Korean], *Journal of Tourism Sciences Society of Korea*, 16: 209–18.

Kang, S.K.-M. and Page, S.J. (2000) 'Tourism, migration and emigration: travel patterns of Korean–New Zealanders in the 1990s', *Tourism Geographies*, 2: 50–65.

Kennett, B. (2002) 'Language learners as cultural tourists', *Annals of Tourism Research*, 29(2): 557–59.

Kim, S.T. (2001) 'A study on supporting measures for improving the travel agency industry' [in Korean], Research Report, Seoul: Korea Culture and Tourism Policy Institute (KCTPI).

Kloosterman, R. (2000) 'Immigrant entrepreneurship and the institutional context: a theoretical exploration', in J. Rath (ed.) *Immigrant Businesses: the economic, political and social environment*, Basingstoke/New York: Macmillan/St Martin's Press.

KNTO (Korean National Tourism Organization) (2005) 'Roles and symbol', available HTTP: <www.knto.or.kr/eng/02_aboutknto/02_01.html> (accessed 1 March 2005).

Langlois, A. and Razin, E. (1989) 'Self-employment among ethnic minorities in Canadian metropolitan areas', *Canadian Journal of Regional Science*, 12: 335–54.

Li, P. (1994) 'Self-employment and its economic return for visible minorities in Canada', in D.M. Saunders (ed.) *New Approaches to Employment Management, Volume 2: discrimination in employment*, Greenwich CT: Jai Press.

Li, P. (2000) 'Economic returns of immigrants' self-employment', *Canadian Journal of Sociology*, 25: 1–34.

Li, W., Dymski, G., Zhou, Y., Aladaba, C. and Chee, M. (2002) 'Chinese American banking and community in Los Angeles County: the financial sector and Chinatown/ethnoburb development', *Annals of the Association of American Geographers*, 92: 777–96.

Light, I. and Bonacich, E. (1988) *Immigrant Entrepreneurs: Koreans in Los Angeles, 1965–1982*, Los Angeles: University of California Press.

Lo, L. (forthcoming) 'The new Chinese business sector in Toronto: a spatial and structural anatomy of medium- and large-sized firms', in E. Fong and C. Luk (eds) *Chinese Ethnic Economy: global and local perspectives*, Philadelphia: Temple University Press.

MEHRD (Ministry of Education and Human Resources Development, Seoul) (2002a) 'Statistics on Korean K-12 students overseas' [in Korean], available HTTP: <www.moe.go.kr/bbs/board.php?bt=data&db=bbs1_7&limit=10&catmenu=m05_02_11&tn=moebbs&action=view&id=69&page=4> (accessed 18 November 2003).

MEHRD (Ministry of Education and Human Resources Development, Seoul) (2002b) 'Statistics on Korean post-secondary students overseas', [in Korean], available HTTP: <www.moe.go.kr/bbs/board.php?bt=data&db=bbs1_7&limit=10&catmenu=m05_02_11&tn=moebbs&action=view&id=63&page=4> (accessed 18 November 2003).

Nash, J. (2000) 'Global integration and the commodification of Culture', *Ethnology*, 39(2): 129–31.

Oakes, T.S. (1993) 'The cultural space of modernity: ethnic tourism and place identity in China', *Environment and Planning D: Society and Space*, 11: 47–66.

Ohmae, K. (1993) 'The rise of region-state', *Foreign Affairs*, 72(2): 78–87.

Ong, A. (1999) *Flexible Citizenship: the cultural logic of transnationality*, Durham: Duke University Press.

Phizacklea, A. and Ram, M. (1996) 'Being your own boss: ethnic minority entrepreneurs in comparative perspective', *Work, Employment and Society*, 10: 319–39.

Portes, A., Haller, W.J. and Guarnizo, L.E. (2002) 'Transnational entrepreneurs: an alternative form of immigrant economic adaptation', *American Sociological Review*, 67: 278–98.

Rath, J. (ed.) (2002) *Unravelling the Rag Trade: immigrant entrepreneurship in seven world cities*. Oxford and New York: Berg.

Ries, J. and Head, K. (1998) 'Immigration and trade creation: econometric evidence from Canada', *Canadian Journal of Economics*, 31: 47–62.

Teixeira, C. (1998) 'Cultural resources and ethnic entrepreneurship: a case study of the Portuguese real estate industry in Toronto', *The Canadian Geographer*, 42: 267–81.

Tourism British Columbia (2001) 'Tourism indicators, December 2001', available http://www.tourismbc.com/PDF/tourism_indicators24.pdf (accessed 1 March 2005).

Tourism British Columbia (2002) 'Tourism indicators, December 2002', available http://www.tourismbc.com/PDF/tourism_indicators31.pdf (accessed 1 March 2005).

VEDC (Vancouver Economic Development Commission) (2003) *Vancouver's English Language School Sector*, Vancouver: VEDC.

Vertovec, S. (1999) 'Conceiving and researching transnationalism', *Ethnic and Racial Studies*, 22: 447–62.

Waldinger, R. and Lichter, M.I. (2003) *How the Other Half Works: immigration and the social division of labor*, Berkeley: University of California Press.

Walton-Roberts, M. and Hiebert, D. (1997) 'Immigration, entrepreneurship and the family: Indo-Canadian enterprise in the construction industry of Greater Vancouver', *Canadian Journal of Regional Science*, 20: 119–40.

Wang, N. (1999) 'Rethinking authenticity in tourism experience', *Annals of Tourism Research*, 26(2): 349–70.

Williams, A.M. and Hall, C.M. (2000) 'Tourism and migration: new relationships between production and consumption', *Tourism Geographies*, 2(1): 5–27.

Yeung, H. and Olds, K. (eds) (2000) *The Globalization of Chinese Business Firms*, Basingstoke: Macmillan.

Zhou, Y. (1998) 'Beyond enclaves: location strategies of Chinese producer service firms in Los Angeles', *Economic Geography*, 74: 228–51.

Zukin, S. (1990) 'Socio-spatial prototypes of a new organization of consumption: the role of real cultural capital', *Sociology*, 24(1): 37–56.

3 Urban boosterism, tourism and ethnic minority enterprise in Birmingham

Trevor Jones and Monder Ram

People don't need to be given their cultures, only their rights.

(Sivanandan 2001)

Like many another old industrial centre, England's self-proclaimed 'second city' is engaged in a process of self-reinvention. In the new age of entrepreneurial local government – with each city striving to attract its due share of inward investment and tourist spending – Birmingham is attempting to make the surreal transformation from nuts-and-bolts-making to post-modernist hedonism. Such a makeover is more than usually difficult in this case, since this is a city whose image, elsewhere in the UK certainly, is emphatically negative. Moreover, it is negative in the worst possible way, for Birmingham is perceived as boring, a flat colourless amalgam of nothing particularly memorable, attractive or distinctive, a world of grey concrete canyons and high-rise social housing, a monument to regimented (and now obsolete) Fordism. Even the local 'Brummie' accent is ranked as the least popular of all the English regional dialects (Webster 2001). Clearly this last point is highly subjective, especially as one of the present authors actually speaks it. But subjectivity is, of course, the very name of the place promotion game. In the great interurban struggle for pleasure-consumers and spectacle-gazers, we might assume a place called Boresville might be last in the queue.

As well as radically transforming the skyline and built environment and providing multifarious leisure attractions, the city government is also engaged in a project of cultural re-imaging. Highlighting the presence of large, Third-World-origin ethnic minorities, the thrust of the campaign is to present the city as a vibrant hub of multicultural diversity, thereby overriding the deep-rooted perceptions of urban monotony. The pivotal resource here is ethnic cuisine – purveyed by the city's innumerable Indian, Pakistani, Bangladeshi, Chinese and African Caribbean restaurants and takeaway outlets – which local decision-makers envisage both as direct magnets for visitors and as enriching the experience of those who visit other attractions. In this chapter we take a somewhat puritanical, not to say Luddite, view of all this, by switching the focus away from the fashionable obsession with the urban as pleasurable consumption (Mullins 1991) towards the people who perform the actual *production* of what is consumed. Drawing from our own previous in-depth surveys of the South Asian restaurant industry in Birmingham (Ram *et al.* 2002), we argue that those who enjoy the great Birmingham multicultural experience do so on the backs of some of its most socially excluded citizens (some of whom do not even officially 'exist'). These are the residents who are least likely to have a stake in the fruits of this great urban rebranding exercise. In the course of our exploration, we also necessarily raise questions about the authenticity of the consumer's experience and about the nature of multiculturalism itself. First, however, we need some context.

Tourism and the reinvention of the second city

Before the 1980s, tourism studies were very much concerned with the international tourist, who tended to be portrayed as a key player in a Gunder Frankian global core–periphery drama, acting as an unwitting agent of 'neocolonialism and social, cultural and ecological mischief' (Graburn and Barthel-Bouchier 2001: 147). Economically, tourism operated as a means of exploiting cheap local labour and repatriating profit back to the core-based consortia that owned it; on the cultural and environmental levels, its invasive impact hardly needs rehearsing. In this version of global tourism, local people are required to act as the 'household staff of global capitalists' (MacCannell 1992: 5), and the spotlight continues to illuminate the growing collateral damage inflicted by this most intrusive, and indeed child-abusing, of industries (Castells 2000). More recently, however, many writers in search of balance have shifted the focus to the tourist as a conscious agent, with an independent agenda often very much at odds with the prevailing logic of neocolonialist capitalism. While the latter logic may well continue to apply to mass-produced package tourism, where consumers are motivated by a combination of hedonism and cheapness, writers like MacCannell (1988) choose to focus on middle-class urbanites in search of 'authenticity', and hence more interested in conserving places than simply consuming them (Urry 1995). Yet local authenticity is hard to come by, and even these dedicated 'posttourists' are more likely to find a commercialized caricature of it, a staged and contrived fake spectacle (MacCannell 1988; Perkins and Thorns 2001). This ersatz factor should be borne in mind as we discuss the ethnic dining-out experience.

The changing ways of looking at tourism partly reflect the ways in which tourism itself is changing in response to the post-industrial shift. Although the mass tourism expressive of the Fordist regime has far from passed away, there are signs of a progressive switch towards flexi-tourism, with the promotion of opportunities for more individualized leisure experiences often incorporating a wider range of visitor destinations (Perkins and Thorns 2001). As well as the continuing international flow from core to periphery, the major cities of the core itself have for some time been busily engaged in the promotion of a new tourist economy to substitute for jobs lost through deindustrialization and the collapse of the old economy (Harvey 1987). Indeed, Mullins (1991) argues that the 'consumption of pleasure' is now the dominant urban theme, superseding traditional urban functions like the production of necessary goods and services. Much in the same vein, Scott (2000) talks of the urban 'cultural economy', a cluster of activities including fashion, media, books and other objects of sensual, mental and emotional gratification which, as Scott notes, are now increasingly produced as commodities. Historically such goods and services were defined as luxuries, and until well into the twentieth century they were virtually the exclusive preserve of social elites. This would apply equally to travel and 'place consumption', and, significantly, tourist services comprise an integral part of Scott's cultural economy.

In this new pleasure- and culture-driven urban context, the successful city is one that is perceived as exciting, stimulating and uplifting, both visually and experientially. According to Perkins and Thorns (2001: 189), 'Place promotion has become a major industry in numerous cities round the world as city boosters attempt to … competitively attract visitors and therefore money into the local economy'. In this interurban scramble to pull in visitors and other bearers of job and income generation, image has become more important than substance, with city leaders striving to conjure up 'evocative symbols to convince investors, consumers, potential residents etc that their city is vibrant, innovative, fun, profitable and a good place in which to live' (Mullins 1991: 330). For Harvey (1990), the actual differences between places are being ironed out under the homogenizing influence of globalization, so they must be replaced by symbolic differences, 'real virtuality' as Castells (2000) calls it. New place identities have to be created which pander directly to the 'tourist gaze' (Urry 1990), and 'a battle occurs over the images of places, that they might appear attractive and desirable' (Savage and Warde 1993: 129).

Given the obvious role of the built environment in sending out visual signals, it is no surprise that place promotion almost always involves physical redevelopment first and foremost. This requirement acquires an extra edge in an age infused with the notion of the city 'as a theatre, a spectacle full of play spaces' (Harvey 1987: 284). Even more in the case of old industrial cities – where a switch from producing unsexy material objects to consuming pleasure requires nothing less than an absolute role reversal – a physical makeover is a mandatory exercise, tangible evidence of the city's magical conversion from sow's ear to silk purse. As well as introducing conspicuous symbols of contemporaneity and future orientation, physical redevelopment also enables a rewriting (even a complete deletion, where required) of the past through selective conservation and obliteration. Thus, in Birmingham, we might argue that redevelopment (see Webster 2001 for an extensive account) essentially consists of three overlapping projects: (a) the implantation of a new twenty-first-century skyline in the form of great flagship projects such as Arena Central (a complex of hotel, offices and residential units planned as Britain's highest building) which will enable Birmingham to 'take its place among the world's great landmark cities' (Webster 2001: 37); (b) the elimination of much of its recent twentieth-century past as Britain's unloved motown and maker of widgets; (c) the rediscovery, rejuvenation and re-branding of selected items of its pre-twentieth-century heritage, in line with Harvey's (1987: 277) sardonic observation that 'the recuperation of "history" and "community" become essential selling gimmicks to the producer of built environments'.

As an example of (b), extensive pedestrianization schemes are gradually replacing the grim flyovers and underpasses of the era of undisputed car domination, while consigned to terminal oblivion is the Bull Ring centre, an archetypal 1960s modernist concrete monster. Arguably the process of destruction is at least as important as that of creation in a city whose major image problem is that of dullness in a world where to be 'boring' is the kiss of death (Webster 2001). Indeed, Birmingham's image is perfectly captured by the fictional name 'Rummidge', affectionately bestowed upon it by its resident novelist David Lodge, a name of grinding tedium that drops off the tongue like a double dull thud. To be strictly accurate, during its early postwar car-making and metal-bashing phase, Birmingham enjoyed a somewhat enviable reputation as a hive of prosperity and high working-class wages, 'one of the most affluent and fully employed cities in Britain' (Rex and Moore 1968: 20). Even so, these virtues are emphatically materialistic, prosaic and lacking soul, hardly bearing comparison with a city like Liverpool – down at heel, physically and economically wrecked, a battered old punch-drunk fighter of a city, but still the place that changed the face of rock'n'roll, and with an instantly recognizable skyline and world-famous football club. Worse still, since deindustrialization, even Birmingham's carpet of affluence has been pulled from under it, reducing it to the status of unemployment black spot. As Merrifield (2000) rightly points out, urban dystopia on the Liverpudlian model can perversely achieve a turn-on in the way that a gangster is always more charismatic than a chartered accountant. But Birmingham, if its detractors are to be believed, is dystopic in an utterly sterile and undramatic way. Nowhere is this unhipness more blatant than in the robotic grey utilitarianism of its early postwar-built environment. So, as the Weakest Link, this must go. Goodbye!

Yet, even as this particular historical slate is wiped clean, others are being redrawn under (c) as 'sites of collective memory' (Harvey 1987: 284). As a reminder of the city's pre-Fordist industrial heritage as a centre of high-value artisanry, the old Jewellery Quarter 'has been repackaged largely for the tourist market' (Webster 2001: 36). A strategic new addition to the city's cultural economy, this revamped jewel zone is immediately evocative of exclusiveness, style, self-indulgence and general naughtiness, an ingenious but oh-so-simple post-modernist reprise of a mystical pre-modern craft. Similarly benefiting from the prevailing fashion for retro is the extensive local canal system, a key element in the local economy from late eighteenth century, until it was superseded by the railways from the 1840s, and gradually fell into abject dereliction. Now cleaned up and redeveloped as a shared space for walkers, boaters, fishers, cafe-goers and wildlife, it forms 'an integral and enhancing feature of the redevelopment of central Birmingham' (Iafrati 2000: 29). As Iafrati also notes, all this green quasi-gentrification chimes perfectly with an overarching theme of 'from labour to leisure', the very essence of the post-industrial shift – or so we are told. While the oft-repeated boast that Birmingham has more canals than Venice was once no more than self-deprecating black humour, it might now strike a faint chord of credibility, especially as these newly sanitized waterways, purged of their rusting bedsteads and dead dogs, are likely to be rather less smelly today than their ancient, rotting Venetian competitors.

Multiculturalism and ethnic cuisine

Seamlessly inserted into this variegated, but broadly coherent, collage is the self-promotion of Birmingham as a site of multiculturalism. Places are often defined and characterized without direct reference to their physical attributes (Harvey 1990), and we might think that the human life that goes on within a place offers even more possibilities for cosmetic surgery than does the built façade. Certainly any city that can credibly project itself as cosmopolitan, culturally diverse and even 'exotic' possesses a useful weapon in the great interurban beauty contest. Accordingly, 'leaders in Birmingham have felt able to seize upon cultural diversity as a bonus, for its exotic appeal, for its cuisine, for its multiplicities of ways of life in one location' (Webster 2001: 38). As is well known, Birmingham was a major destination for the great postwar labour migrant streams from the Caribbean, the Indian subcontinent and to a lesser extent Hong Kong and, not coincidentally, it provided the setting for Rex and Moore's (1968) pioneering study of racism and racialized minorities. By the 1991 census, the South Asian-origin population (Bangladeshi, Indian, Pakistani) of Rex and Moore's study area, the inner-city ward Sparkbrook, had exceeded the 50 per cent mark (Ram *et al.* 2002), and it is expected that within the next two decades Birmingham as a whole will attain this 'majority–minority' status (Slater 1996).

Lest we forget that this presentation of ethnic minorities as a positive civic resource constitutes in itself a most monstrous re-branding exercise, a cultural U-turn requiring historical amnesia on an heroic scale, we may care to remind ourselves that during the initial era of mass immigration this city and its region became notorious as an epicentre of popular racism. Far from a valuable human resource, immigrants were consistently patholo-gized and problematized as an entirely negative quantity. This is succinctly captured by Rex and Moore (1968: 19–20), who remind us of the manner in which 'racial problems dominated public discussion' in the early 1960s, and who lament that 'often the sentiments expressed by those who wrote to the [local] newspapers ... smacked more of the Deep South in the United States or of settler Africa than of the city of Reform'. Moreover, at that very time, one of the MPs for neighbouring Wolverhampton, J. Enoch Powell, was busily acquiring notoriety as the leading parliamentary voice of the anti-immigration lobby, the official legitimizer of popular racism.

The rewriting of history is by no means confined to Birmingham and the West Midlands. Recently, Salil Tripath (2003) has written of the 'collective amnesia' descending on the city of Leicester on the 30th anniversary of the arrival of the first expelled Ugandan Asians. In stark contrast to the naked panic expressed by local government officials at the time, this entrepreneurial ethnic community is now officially celebrated for its creation of 30,000 jobs and its contribution to the new multicultural Leicester. Indeed, such mental contortions remind us that the very concept of multiculturalism is itself highly problematic – seen by many insiders as a diversionary smokescreen to obscure the fight against racism. For Kundnani (2002: 68), multiculturalism essentially provides a means of social control, with minority cultures 'turned from a living movement into an object of passive contemplation'. All this must be seen as part of the rise of identity politics, which, according to critics like Kundnani and Sivanandan (2001), swamps the acknowledgement of racism by insisting first and foremost on cultural recognition. Any analysis of the urban-tourism-through-ethnic-enterprise project is bound to be haunted by these kinds of question marks.

This is not to dismiss such an approach as entirely negative and valueless. Though it is always dangerous to try to bury the past – there is no disposal service for the Dustbin of History – we shall take the charitable view here that, at the very least, these local authorities are now sending out the right signals. In itself, this is a far cry from the 1960s stance, when Birmingham's bureaucratic allocation procedures were identified by Rex and Moore as exacerbating racist conflicts in the local housing market, the key bone of contention. Historical about-turns are not necessarily bad in themselves if they are genuinely intended to exorcize a regrettable past, and so rare are positive official portrayals (however belated) of racialized minorities that they should perhaps be warmly welcomed – though not of course unconditionally. In any case, to adopt an uncompromising position of absolute purity is to rule out the possibility of any positive initiative whatsoever.

Returning to the technicalities, it is evident that, at the rudimentary demographic level, Birmingham contains a very substantial and growing racialized minority population, and hence the 'raw material' with which to offer itself to the tourist gaze as a place of great human diversity, a veritable melting pot of the peoples of the Earth. Such intermixing of ethnicities, languages, religions and lifestyles is far removed from the grey monochrome of its traditional image, indeed a direct denial of it. Moreover, all this is by no means pure, abstract narrative. With regard to visible urban imagery, there is no doubt that the presence of these large ethnic minority communities has had a transformative effect, not least at the level of urban spectacle. Especially salient here is the heavy involvement of ethnic minorities in entrepreneurial small business activities, notably in retailing and catering. On a symbolic level, the mere presence of these clusters of 'exotic' shops and restaurants helps to breathe life into urban neighbourhoods, not only adding visual colour but also reproducing the classic urban milieu as envisaged by Jane Jacobs (1961) – the densely packed commingling of diverse land uses in the same space (Merrifield 2000). Complementing this social animation is the sheer sensuality of nostrils twitching to the scent of spicy cooking, eyes boggling at the dazzling sari shop displays, hips grinding to the sounds of bangra and reggae.

At a more tangible level, ethnic restaurants have a direct role to play in the new pleasure consumption economy. Not only can they act as attractions for visitors in their own right, but they can also provide a dining out experience for those pulled in by other attractions such as the National Exhibition Centre. This potential has been officially recognized by the incorporation of Chinatown into Birmingham's Entertainment Zone (Webster 2001) – though probably even greater attractive potential lies within the South Asian curry house sector, whose reputation already extends far beyond the city's bounds. Prominent here is the so-called Balti Quarter, a highly localized inner-city cluster of some 60 South Asian restaurants enjoying a combined annual turnover in 1999 of around £8.5 million (€14 million; Ram *et al.* 2002). The great bulk of this customer base is drawn from the mainstream white population, predominantly middle-class people and students.

A textbook example of self-promotion, the Balti Quarter is well known outside Birmingham to connoisseurs of the cuisine, and several of our own interview respondents report that they pull in custom from as far away as Wales and North-West England, while others are frequently patronized by foreign visitors (Ram *et al.* 2002). From this it would seem that the Balti Quarter is well ahead of the place promotion game, irrespective of any intervention by local government, and already possesses much of the necessary infrastructure to accommodate whatever tourist role the city leaders may have in mind for it. With the faintest touch of irony, we might suggest that, for those lacking the resources for a Third World holiday, there is always the alternative of a short break in Birmingham's Balti Quarter. With the publicly subsidized addition of the appropriate signifiers – perhaps a few entrance arches in the shape of the Taj Mahal, a giant electronic billboard with flashing pink elephants and characters from Disney's *Jungle Book*, piped eezi-listen sitar music, some amputees paid by Birmingham City Council to sit around in rags with begging bowls, some strategically scattered dollops of hygienic washable non-odorous plastic cow dung – a total genuine fake 'Indian village' environment could be created, which would be realer than the real thing in a world where there no longer *is* a 'real thing'. And, even more to the point, it would not contravene local public health bylaws.

Mention of the 'real thing' reminds us that the entire 'Indian' restaurant phenomenon is itself riddled with virtuality and quasification (to use the terminology of Castells 2000). In the first place, many so-called Indian restaurants are owned by Pakistanis, or even more often by Bangladeshis who traditionally are unusually concentrated in the catering trade (though presumably not when they live in Bangladesh, but this is by the by). Even so, the label always takes priority over the actuality, and so all that matters is that 'Indian' is now popularly accepted shorthand for all cuisine deriving from that vast subcontinent, with its infinite range of nationalities, subnationalities and culinary styles. Moreover, as Narayan (1995) reminds us, that sine qua non of the entire cuisine, curry, is itself a bastardized colonialist appropriation from the days of the British Raj, a repackaging of ingredients in a way palatable to Europeans, even though not hitherto cooked or eaten in that form by the indigenous colonized peoples. Apparently real virtuality had already set in by the Victorian High Noon of the British Empire but, since Derrida and co. had yet to be born, no one noticed it at the time.

More to the immediate point, what could be more appropriate than that an invented culinary authenticity be appropriated by a programme of invented urban authenticity? The metaphorical symmetry here is little short of awesome. But the relentless march of re-authentification does not end here. Indeed, in the specific instance of balti cuisine, the culinary re-branding has been taken a stage further in that, a decade or so ago, an enterprising Birmingham restaurateur (one of our very own respondents claims the distinction) decided to fine-tune certain existing, well established curry recipes and offer them under this brand new title (Ram *et al.* 2002). With a host of imitators rushing to follow this innovative mode of product differentiation, in no time a whole eating-house district succeeded in finding a novel identity for itself and projecting this almost nationwide.

Here we would hasten to add, firstly, that both the present authors are gluttons for the stuff, so there is no problem with our 'situatedness'; and, secondly, that no deception of any kind is being practised here – nor is that indeed possible in a world where prosaic matters like where the food comes from, or what it tastes like or costs, have ceased to register. In this respect, food tourism is no different from any other tourism. Eating out, as Beardsworth and Bryman (1999) assure us, is all about interactive theatre, with diners as well as waiters, chef and maitre d' all active members of the cast (and the kitchen porters presumably the stage hands who receive no curtain call and not much pay either, as we shall see; oh, by the way, who is the audience here?).

There is a clear marketing opportunity here for an enterprising operator to capture the whole procedure in a video game, which would save everyone the trouble of actually going out. Eating out is also concerned with the quest for social distinction and cultural credibility (Warde *et al.* 1999), a Holy Grail best sought after amid the mystique of ethnic and 'exotic' eating houses where – since almost the entire white population is basically ignorant on the matter – any Tom, Dick or Harriet can assume the mantle of connoisseur, arbiter of taste and balti bore (while thee waiters snigger behind their serviettes). The entire exchange is one of complicity and mutual conspiracy, with tacitly observed rules and rituals (Liu and Fine 1995). When all is said and done, the average white punter prefers to be mesmerized by Eastern magic and would be tremendously upset by its demystification.

In case any of the above is misinterpreted as some kind of negative judgment on South Asian restaurateurs themselves, we would clarify that if any kind of judgmentalism is operating here, it is directed entirely at the surrealist context in which they find themselves, the perverse desires to which they are obliged to pander, and the highly stressful market forces with which they must do daily battle. In practice, the great major-ity of our own respondents (Ram *et al.* 2002) are dedicated exemplars of the craft ethic, motivated more by the compliments paid to them by their customers than by the profits and growth which neoclassical microeco-nomics tells us are the driving forces of enterprise. This is beautifully encapsulated by one of our Indian respondents: 'The biggest kick I get is when people compliment, "Oh my goodness, I can't get this food in India", and in this he speaks for many others. Furthermore, not a few of our interviewees have strategically distanced themselves from the balti bandwagon, either literally by locating well away from Baltiland itself or by conscious product differentiation. As an example of the latter, we could do no better than the Indian Sikh proprietor who asked himself, 'Why not go back to the roots? We only do traditional Punjabi dishes and anybody who comes through that door and comes out with a fancy name, we tell them we don't do it'. Like others, he scorns balti as a manufactured concept, a kind of mcdonaldization of the true wondrous gastronomy of the subcontinent. Revealingly, he estimates his custom to be around 90 per cent Indian, so it would seem that this genuinely genuine ethnic eating experience has escaped the notice of all but the most discerning white seekers-after-social-distinction.

Multicultural food tourism: the costs

Entrepreneurs

On the face of it, the curry house proprietor in present-day Britain would seem to enjoy an enviably strong market position, unchallenged king (there are very few female owners, so this is not a gender-blind usage) of an appar-ently insatiable market, growing at over 10 per cent a year in the late 1990s

(Ram *et al.* 2002). The contrast here with the bulk of UK Asian businesses – located as they are in overcrowded, declining sectors like small food retailing – is absolute. Not only is the curry sector growing vigorously, but Asian restaurateurs ostensibly enjoy a considerable degree of market leverage, in that they are protected from competition from non-Asians, who for obvious reasons lack credibility as curry house proprietors. Even more so in a city like Birmingham, extra strength seems promised via the promotion of multicultural tourism, where (it is to be presumed) the local state will be throwing its weight behind its chosen dramatis personae – the Asian restaurateurs and their employees – providing all manner of infrastructural, financial, technical and other forms of support to ensure the realization of its multicultural vision.

In practice, however, this rose-tinted scenario is far from the present experience of many if not most Asian restaurateurs, and it is unlikely to be realized in the foreseeable future without a truly immense market expansion. Not untypical of the problem is one of our Indian respondents who, though long established, is struggling to make ends meet and surviving only on the back of his wife's paid employment. He laments, 'I am keeping afloat by working seven days a week, that is what I've been doing for 17 years, only one day off a year Christmas Day. I don't want the children to get involved because I don't think this is a good life'. Speaking for the trade as a whole, a Bangladeshi owner estimates that 'nowadays you find 10 per cent of people can make it big, 40 per cent just make a living and 50 per cent are struggling'. Note here that even bare survival is only possible by working long and hard into the small hours, with owners and workers alike unlikely to make it to bed much before 2 a.m. According to another respondent, this time Pakistani, 'I think you are better off just working for somebody else'. Others, equally disillusioned, claim they would be better off 'just sitting at home', in the words of one.

Across the entire restaurateur sample, intensification of competition was repeatedly identified as the primary cause of this gruelling and under-rewarded toil (Ram *et al.* 2000). As we have argued elsewhere (Ram *et al.* 2002), the restaurant sector is no less subject to market saturation than most of the other sectors in which Asian businesses are concentrated, a paradoxical situation at first sight. Unlike retailing and CTN (confectionery–news–tobacco), where the impact of supermarket and other corporate competition has wrought catastrophic havoc on the small Asian entrepreneur's market position (Jones and Ram 2003), the Asian restaurateur basks in a sector with explosive growth in customer spending, underpinned by a seemingly inexhaustible white middle-class consumer base. Nevertheless, the reality is that restaurants, takeouts and other supply outlets have grown even faster than demand – the result of what might be described as a 'gold-rush effect', where excessively large numbers of prospectors are drawn in by the hope of rich pickings. As in an actual goldfield, it is generally the first entrants who thrive, while the later arrivals struggle.

Enthusiasts for the benign qualities of entrepreneurialism as liberator and enricher of the individual are fond of extolling the profit-making opportunities offered by the various 'niche markets' supposedly opened up by the post-industrial economy. By their very nature, however, such niches tend to fill up very rapidly and, unhappily, South Asian business owners in Britain have discovered that every niche they have pioneered has soon been flooded by a surfeit of competitors – fellow Asians metaphorically slitting one another's throats (Jones *et al.* 2000). Even that most niche-like of niches, exotic catering, has not proved immune from the process, despite Asians enjoying both customer growth and an effective ethnic monopoly. While competition among the balti houses themselves constitutes an onerous threat in its own right, owners also identified additional competitive pressures from non-balti outlets like fish-and-chip and fast-food shops and from the growth of balti houses in other parts of the city region (Ram *et al.* 2000).

At this juncture, we need to qualify any impression that these restaurant owners are no more than pawn-like victims of an antagonistic commercial environment. Indeed this is one of the pivotal issues in the entire debate about ethnic minority enterprise: the question of self-determination versus external determination, agency versus structure. For our own part, we have consistently laid stress on the political–economic environment in which such entrepreneurs are embedded, a stress designed to counter over-optimistic accounts of Asian business as a success story resting on the autonomous resources of the ethnic community. Nevertheless, there is still a sense in which, like any other social actor, Asian entrepreneurs must be recognized as active agents exerting free will and choice to adapt to market pressures which in themselves are far from absolutely inevitable or deterministic. On the contrary, there is a range of managerial options – ostensibly at least – and Asian restaurateurs can be seen to be striving with considerable ingenuity to ameliorate and adapt to competitive stress (see Ram *et al.* 2002 for fuller account of adaptive strategies). Most successful are those who have 'gone upmarket', differentiating themselves from the competition by the conspicuous opulence of their premises and the reputation of their cuisine. Often such businesses are (re)located in affluent suburban or city centre premises well removed from the inner-city squalor of the Balti Quarter, distancing themselves physically as well as socially from the mass. Yet in the last instance, however proactive and creative their strategy, Asian restaurateurs' strategic space is rigorously constrained within parameters fixed by external market conditions. Success is strictly rationed by price in the capitalist market, and because the upmarket strategy is essentially capital-intensive, it is open only to a minority of small Asian entrepreneurs. Typically, Asian entrepreneurs are under-capitalized and lack the means to execute ambitious projects requiring costly sites, premises and equipment (Ram *et al.* 2002). Proactive adaptation is certainly to be recommended, but it cannot be piously assumed, given the average entrepreneur's non-possession of the necessary resources.

If this line of reasoning still appears to play down the force of entrepreneurial agency, we would remind the reader that we are far from alone in questioning the small business owner's freedom of self-determination. Chapman (1999) is one observer who disputes whether the term 'strategy' is applicable at all in small firms, where the owner's workload is too time-consuming to allow for strategic thinking, and where management is more likely to consist of *ad hoc* reaction, 'just muddling through on the basis of custom, trial and error; ... a course of action is not a strategy if it has been forced on one by circumstance' (Chapman 1999: 77). In similar vein, Curran *et al.* (1991) refer to small business strategies as 'hypothetical'. Significantly, these writers are addressing small firms as a whole rather than those owned by ethnic minorities, whom we might suppose to be even more constrained.

In the light of these strategic non-options, it comes as no surprise that Birmingham Asian restaurateurs' responses to market pressures tend to take the familiar low-tech labour-intensive cost-cutting form that has bedevilled Asian business owners in Britain from the very outset. As the case of the Punjabi restaurant cited above suggests, this need not always be disadvantageous, since an effective low-cost strategy is available through the creation of a unique gastronomic identity. Yet even this depends on high-quality culinary skills, another scarce resource beyond the reach of many. More typically, restaurateurs rely for their survival on a very basic form of price competition, with menu tariffs maintained at a low level, dependent on the entrepreneur's willingness to accept uneconomic returns for long and unsocial working hours. While this is certainly to be seen as active agency, it is hardly an inspiring recommendation for the virtues of free will and choice – nor indeed for multicultural urban reinvention. To put it baldly, Birmingham's vaunted multicultural culinary experience, its vast choice of innumerable and affordable curry houses, would be decisively compromised if all Asian restaurateurs operated according to capitalist calculative rationality and actually insisted on profitability as a condition for their existence. If, on top of this, they were also to desist from cutting regulatory corners, the outcome could only be a drastic reduction in the number of catering outlets. The sector's continuance and further development as an urban re-branding exercise rests entirely on Asian entrepreneurial self-exploitation (not to mention self-criminalization) – and indeed on worker exploitation, as we shall discover in the following section.

Employees

Running incessantly throughout the discourse in this field is the theme of the family as a decisive source of labour power for the ethnic minority business. Whether in catering or elsewhere in the economy, family members are seen as a trusted, flexible and cut-price labour supply that gives these businesses a competitive edge (see Ram and Jones 1998 for a review). Highly advantageous in principle, in practice family business units can be riddled with all manner of schisms on generational and gender lines and can be hampered by nepotistic loyalties (Ram and Jones 1998). Moreover, in the case of restaurants, the role of family members is qualified by two special considerations. First, the rising expectations of the British-born generation – and often of parents on behalf of their children – mean that fewer young family members wish, or are compelled, to submit to the drudgery of the late-night catering business. Second, the necessary scale and division of labour in the typical restaurant compels additional recruitment beyond the familial bounds, though almost always within the co-ethnic community. Scholarly accounts of the resulting boss–worker relationships have tended to be polarized between the narra-tive of sweated exploitation recounted by writers like Mitter (1986) and the idealistic representations of workplace harmony, where face-to-face paternalism and common ethnic bonds can apparently transcend the normal class antagonisms (Werbner 1990).

When these ethnic minority models of industrial relations are set against the mainstream discourse on small firms, one is struck by the degree of similarity – by the way in which the behaviour of ethnic minority firms is no more than a variation on a universal theme. According to Chapman (1999), the belief in small-firm working relationships as essentially harmo-nious is a long-standing theme applying right across the board, irrespective of ethnicity. Over time, the harmony thesis has become subject to critical scrutiny, with the small workplace increasingly recognized as the site of a complex range of possibilities. One possibility is that control is exerted through persuading workers to collude willingly in the process – though, as Goss (1991) makes clear, there are various degrees of this. His own typology of small-firm control strategies ranges from fraternalism and paternalism, where the boss's authority is heavily disguised under a veil of personalized face-to-face pseudo-equality; through benevolent autocracy, where authority is overt but tempered by grace and favour; to sweating, which speaks for itself. In applying this kind of mainstream thinking to Asian firms, Ram (1994) argues for what he describes as 'negotiated pater-nalism', a regime far from conflict-free, but in which employees enjoy more than the customary leverage.

Certainly many of our own interviews bear testimony to the benefits for workers that lie in these informal industrial relations practices. Among the fringe benefits mentioned are lifts to and from work and free food on the premises, though like any other paternalistic rewards these are contingent on the whim of the employer. Looking more closely at Asian restaurant practices in the light of Goss (1991), we emphasize that relationships are heavily influenced by the degree to which employ-ers are dependent on their workforce. Given the dwindling supply of potential workers and the difficulties of recruitment, it is likely that key workers, such as chefs, stand to gain substantial benefits from a reign of paternalism. Conversely, illegal immigrant kitchen porters (see below) – dependent as they are on their employer for their very presence in Britain – are more likely to be subject to benevolent autocracy at best and sweating at worst.

Whatever the leverage enjoyed by certain privileged workers, there is no getting blood out of a stone, and hence no realistic means of extracting better wages from cash-strapped employers. What is unarguably the case is that the curry house trade in Birmingham pays truly dire wages: even experienced chefs typically draw only between £200 and £250 (€300–400) a week. For all their scarcity value, these crucial elements in the catering process are unable to command much more than the national minimum wage – presumably a reflection of a dearth of alternative opportunities for them outside the curry house niche. Other employees are even less well remunerated, with waiters typically on £150 and kitchen porters often earning as little as £100. It will not escape notice that, assuming a 40-hour working week, waiters' and porters' wages do not even conform to the national minimum wage, with porters getting little more than half the legal rate. This regulatory bypass is accomplished by creative accounting, as with one of our respondents who explains that 'on the books we have five staff when really there are twelve' (Ram *et al.* 2002).

Once again, no judgmentalism is in order, since the hypercompetition characteristic of the Asian restaurant market ensures that, for many owners, survival is guaranteed only by self-exploitation and worker exploitation. Indeed, a consistent view of Asian entrepreneurs as self-empowered free agents demands that they be *congratulated* for their creative and ingenious survival strategies. Moreover, without their survival, both urban multiculturalism and consumer choice would be seriously threatened – a clearly unthinkable scenario when we bear in mind the sacred-cow status of these institutions. Yet even this is only part of the story. The viability of the trade now depends on even more desperate measures – notably on a growing use of undocumented immigrant labour. Given that young, British-born Asians are decreasingly inclined to work for the meagre wages prevalent in the balti sector, the only recourse is to recruit directly from the subcontinent, where living standards and financial expectations are dramatically lower. Given also that most of such primary immigration is now officially blocked, the illegal route is the only remaining recruitment option. This dilemma is summarized by one of our Bangladeshi respondents, who speaks for many when he declares: 'There are more restaurants now than there are people to work them. If I could get workers who were legal, I'd prefer to'.

In a world of grotesque inequalities, there are, of course, many eager candidates for sweated labour and invisibility in a Birmingham kitchen, an eagerness nicely captured by one of our Bangladeshi respondents: 'There is a good life here, those who have not managed to get to the UK envy those who have' (Ram *et al.* 2002; see also Zlotnik 1999 on the global logic of illegal immigration). In view of this – and many other – respondents' obvious contentment with a £100-a-week wage, it may be tempting to conclude that, far from a form of exploitation, illegal restaurant work actually represents opportunity and social mobility. In our own view, such relativism provides a classic justification of social iniquity, in this instance the creation of an utterly powerless, socially excluded underclass, whose members do not even officially exist (CARF 1997).

Conclusions

This brief account of Asian catering and the reinvention of Birmingham could be read as a case study in the polarized restructuring of the post-industrial city – where top-level gentrification is necessarily accompanied by the rise of an army of helots serving the needs of the super-rich (Sassen 1996). Since Birmingham is not in the front rank of world cities, this effect is not as extreme as it might be. Nevertheless, polarization is taking place, and in this chapter we have seen it reflected in the coexistence of two opposing narratives. At the top is a discourse of urban redevelopment, multiculturalism, spectacle, consumer choice and the pursuit of leisure. At the bottom is a tale of marginal economic survival, unsocial hours and under-rewarded toil under precarious conditions of ever-present risk.

From the evidence presented here, it is hard to avoid the conclusion that the costs of the former are substantially borne by the latter. Despite the official rewriting of history – with the Asian presence in the city now presented as a major asset instead of a burden and a socially destabilizing force – it is hard to resist the conclusion that the racialized division of labour, with all the iniquities therein, continues to exist, though with a new post-industrial gloss. Sweated toil in metal foundries has been abolished and replaced by sweated toil in restaurant kitchens – hardly a Brave New World even by the standards of the post-modernist imagination. Like Birmingham itself, racialized minorities have certainly been re-branded, but such is the disjunction between style and substance that this superficial transformation appears likely to effect little material improvement in their lives. Any development beyond the superficial would require substantial infrastructural investment, as well as a radical rethink about the very conditions of existence in the Balti Quarter. Should this situation somehow be brought within the regulated embrace of the formal economy, in order to improve the material standards of both owners and workers? Presumably this would require a carefully designed, integrated strategy, working in close concert with local ethnic business support agencies. Or would that simply kill off many of the participants, not to mention the anarchic, warm-blooded spontaneity, which at present distinguishes it from the mass-produced corporate conformity of the world around it?

References

Beardsworth, A. and Bryman, A. (1999) 'Late modernity and quasification: the case of the themed restaurant', *Sociological Review*, 47(2): 228–57.

CARF (Campaign Against Racism and Fascism) (1997) 'Commentary', *Race & Class*, 39: 85–95.

Castells, M. (2000) *End of Millennium (The information age: economy, society and culture, volume III)*, Oxford: Blackwell.

Chapman P. (1999) 'Managerial control strategies in small firms', *International Small Business Journal*, 17(2): 75–81.

Curran, J., Blackburn, R. and Woods, A. (1991) 'Profiles of the small firm in the service sector', paper presented at ESRC Centre for Research on Small Service Sector Enterprises, Kingston Business School, Kingston University.

Goss, D. (1991) *Small Business and Society*, London: Routledge.

Graburn, N. and Barthel-Bouchier, D. (2001) 'Relocating the tourist', *International Sociology*, 16(2): 147–58.

Harvey, D. (1987) 'Three myths in search of a reality, in urban studies', *Environment and Planning D, Society and Space*, 5(4): 367–76.

Harvey, D. (1990) *The Condition of Postmodernity*, Oxford: Blackwell.

Iafrati, S. (2000) 'From labour to leisure: the changing role of canals in Birmingham', *The Journal of Regional and Local Studies*, 20(1): 29–39.

Jacobs, J. (1961) *The Life and Death of Great American Cities*, New York: Random House.

Jones, T. and Ram, M. (2003) 'Asian business in retreat?', *Journal of Ethnic and Migration Studies*, 29(3): 485–500.

Jones, T., Barrett, G. and McEvoy, D. (2000) 'Market potential as a decisive influence on the performance of ethnic minority business', in J. Rath (ed.) *Immigrant Businesses: the economic, political and social environment*, Basingstoke/New York: Macmillan/ St Martins Press.

Kundnani, A. (2002) 'The death of multiculturalism', *Race & Class*, 43(4): 67–72.

Liu, S. and Fine, G. (1995) 'The presentation of ethnic authenticity: Chinese food as a social accomplishment', *Sociological Quarterly*, 36(3): 535–53.

MacCannell, D. (1988) *The Tourist: a new theory of the leisure class*, New York: Schocken.

MacCannell, D. (1992) *Empty Meeting Grounds: the tourist papers*, London: Routledge.

Merrifield, A. (2000) 'The dialectics of dystopia: disorder and zero tolerance in the city', *International Journal of Urban and Regional Research*, 24(2): 473–89.

Mullins, P. (1991) 'Tourism urbanization', *International Journal of Urban and Regional Research*, 15(3): 326–42.

Narayan, U. (1995) 'Eating cultures: incorporation, identity and food', *Social Identities*, 1(1): 63–86.

Perkins, H. and Thorns, D. (2001) 'Gazing or performing?: reflections on Urry's Tourist Gaze in the context of contemporary experience in the Antipodes', *International Sociology*, 6(2):185–204.

Ram, M. and Jones, T. (1998) *Ethnic Minorities in Business*, London: Small Business Research Trust.

Ram, M., Abbas, T., Sanghera, B. and Hillin, G. (2000) 'Currying favour with the locals: balti owners and business enclaves', *International Journal of Entrepreneurial Behaviour and Research*, 6(1): 41–55.

Ram, M., Jones, T., Abbas, T. and Sanghera, B. (2002) 'Ethnic minority enterprise in its urban context: South Asian restaurants in Birmingham', *International Journal of Urban and Regional Research*, 26(1): 24–40.

Rex, J. and Moore, R. (1968) *Race, Community and Conflict*, Oxford: Oxford University Press.

Sassen, S. (1996) 'New employment regimes in cities: the impact on immigrant workers', *New Community*, 22(4): 579–94.

Savage, M. and Warde, A. (1993) *Urban Sociology, Capitalism and Modernity*, London: Macmillan.

Scott, A. (2000) 'The cultural economy of Paris', *International Journal of Urban and Regional Research*, 24(3): 554–66.

Sivanandan, A. (2001) 'Poverty is the new black', *Race & Class*, 43(2): 1–5.

Slater, T. (1996) 'Birmingham's Black and South Asian population', in A. Gerrard and T. Slater (eds) *Managing a Conurbation: Birmingham and its region*, Studeley, Warwickshire: Brewin Books.

Tripath, S. (2003) 'Powers of transformation', *Index on Censorship*, 32: 125–31.

Urry, J. (1990) *The Tourist Gaze: leisure and travel in contemporary societies*, London: Sage.

Urry, J. (1995) *Consuming Places*, London: Routledge.

Warde, A., Martens, L. and Olsen, W. (1999) 'Consumption and the problem of variety: cultural omnivorousness, social distinction and dining out', *Sociology*, 33(1): 105–28.

Webster, F. (2001) 'Re-inventing place: Birmingham as an information city?', *City*, 5(1): 27–46.

Zlotnik, H. (1999) 'Trends in international migration since 1965: what existing data reveal', *International Migration*, 37(1), 21–61.

4 Ethnic precincts as contradictory tourist spaces

Jock Collins

Two images flash into the minds of Sydney-siders when they think about Cabramatta – an Asian, largely Vietnamese, suburban precinct some 45 kilometres west of Sydney's Opera House. One image is that promoted by the local government authorities and local entrepreneurs: Cabramatta as 'Asia in one city' – the sights, smells, tastes and sound of Asia. Just listen to the Council's promotional compact disc; it tells you where to go to see the many faces of Cabramatta. The message is: 'Come, see and escape into ethnic diversity'. The other image arises from the front pages of the Sydney tabloid, the soundgrabs of a Sydney radio news broadcast or the flickering television images on the six o'clock news. It shows the arrest of another heroin dealer or the bloody aftermath of a Cabramatta stabbing or shooting. This image is one of Cabramatta as heroin capital of Sydney. The message this time is: 'Keep away, danger, Asian gangs, drugs and guns'.

Despite these contradictions, the local municipal council estimates that some 350,000 people visit Cabramatta each year as tourists, though it does not have information on where these tourists come from. Contrast this with the images that people from the rest of the world have of Sydney: the Sydney Opera House, the Sydney Harbour Bridge, Bondi Beach and the 2000 Olympic Games (definitely) and kangaroos and koalas in Sydney's streets (possibly). Sydney is one of the most cosmopolitan cities in the world today (Collins and Castillo 1998; Connell 2000), with 58 per cent of the population of four million either first- or second-generation immigrants from most reaches of the globe. Despite this, place marketers (such as Tourism New South Wales) have emphasized Sydney's natural environment, its unique flora and fauna, but not, curiously, its cosmopolitan character.

When we think about the relationship between tourism, immigration and ethnic diversity in cosmopolitan cities such as Sydney, the ethnic precincts are a good place to start. Ethnic precincts are key spatial sites – though, importantly, not the only sites – of the ethnic economy in the city. In downtown or suburban parts of the city, 'ethnic precincts' are essentially clusters of ethnic minority or immigrant entrepreneurs in areas that are designated as ethnic precincts by place marketers and government officials.

These are characterized by the presence of a substantial number of immigrant or ethnic entrepreneurs, who line the streets of the precinct selling food, goods or services to many co-ethnics and non-co-ethnics alike. Ethnic precincts come in a number of forms. Most are associated with a single ethnic group: Chinatown, Little Italy, Little Korea, Little Vietnam, Little Turkey and so on. These can be distinguished from suburban multi-ethnic precincts, or 'ethnoburbs' (Li 1998), such as Cabramatta. Ethnic precincts are thus a key site of the commodification and marketing of ethnic diversity in cities of significant immigration. They are a prime focus for the production and consumption of the ethnic economy, a commodification of place and space where the symbolic economy of space (Zukin 1995) is constructed on representations of ethnicity and 'immigrantness'.

This chapter explores the contradictions inherent in the relationship between ethnic precincts, immigrants and tourism, taking three Sydney precincts as a case study. When we study the historical development and contemporary nature of ethnic precincts such as Chinatown, Little Italy and Cabramatta ('Asiatown' or 'Vietnamatta'), some of the links between ethnic precincts and ethnic tourism become apparent. The aims of the chapter are to discuss the contradictions inherent in the ethnic precinct; to examine the central role of ethnic entrepreneurs in giving ethnic precincts their particular character; to look at the role of local and provincial government authorities in shaping the emergence of the ethnic precinct; and, finally, to think about the relationship between immigration, urban and suburban ethnic precincts, and tourism.

The chapter begins by reviewing the interdisciplinary literature relevant to ethnic precincts and giving a brief overview of Sydney and its ethnic precincts. Each of the subsequent sections is devoted to a key contradiction identified in the relationship between immigrant settlement, ethnic precincts and tourism in Sydney. First, there are the contradictions inherent in the competing conceptions of authenticity required to make the ethnic urban tourist experience credible – the *problem of credibility and authenticity* of the ethnic precinct. Second, there is the *problem of legitimation*. Who speaks for a particular ethnic authenticity? How is that ethnicity to be symbolically represented? Who decides, and who controls the 'ethnic voice' in the city within resident ethnic groups of complex diversity? Third, there are the problems of crime and socioeconomic inequality in ethnic precincts, a contradiction reflected in the *problem of control and safety* as a necessary part of the strategy for boosting ethnic tourism in the city.

The ethnic economy, ethnic places and spaces

The conceptual framework for this exploration of the contradictory nature of ethnic precincts in Sydney and their relation to tourism draws on a broad-ranging, interdisciplinary literature – and in particular on studies that address issues of the cultural and symbolic economy (Zukin 1995), international tourism (Suvantola 2002; Urry 2002), the urban and cultural geography of ethnic place and space in the city (Burnley 2001), ethnic economies (Light and Gold 2000), immigrant entrepreneurship (Kloosterman and Rath 2003; Rath 2000; Waldinger *et al.* 1990) and the marketing of ethnicity (Halter 2000).

One of the contradictions of globalization is that local difference and place identity become more important in a globalized world. The urban tourism experience consists not only of a collection of tourist facilities, or real economy experiences, but also of a set of symbolic economy experiences. The latter involves the consumption of signs, symbols, festivals and spectacles used in creating *aestheticized* spaces of entertainment and pleasure. This has led researchers to explore the links between ethnic heritage, cultural diversity and urban tourism as crucial components of the cultural capital of post-industrial society (Kearns and Philo 1993; Lash and Urry 1994). In discussing the 'symbolic economy', Zukin (1995) points to the role of ethnic diversity in shaping place and space, and then relates this to a tendency to commodify cosmopolitan lifestyles and turn them into a vital resource for the prosperity and growth of cities. Another point worth stressing is the important role that different regimes of regulation and governance (national, provincial and local) play in shaping patterns of immigrant entrepreneurship, including spatial patterns, in different countries (Kloosterman and Rath 2003).

The most highly developed interdisciplinary theoretical framework for the study of urban ethnic tourism which draws on these elements is *regulation theory* (Hoffman *et al.* 2003), which has become central to the new tourism literature (Costa and Martinotti 2003). Fainstein *et al.* (2003) explore the way that four types of regulatory frameworks – regulation of visitors to protect the city, regulation of the city for the benefit of the visitors and the tourism industry, regulation of tourism labour markets and regulation of the tourism industry itself – structure relations within the urban tourist milieu and provide a framework for a historical and contemporary comparative analysis of global urban tourism. This regulation theory approach 'places tourism within a complex matrix of economic, political, cultural and spatial interactions and illustrates the interplay of sectors and scales – local, regional, national and international', according to Fainstein *et al.* (2003: 240). The authors also stress the importance of studying the linkages and processes inherent in the city tourism experience, 'without sacrificing the possibility of agency or overlooking the complex role of culture'.

Various writers have drawn attention to the spatial aspect of ethnic and cultural consumption, and particularly to places of food consumption (shopping malls, high streets, eating precincts). Gabaccia (1998: 215), who explores the development of ethnic food and restaurants in the USA, cites another author (Mobray) thus: 'Eating is a form of travel, and no matter how high the price of cardamom, taste bud tourism is a real bargain'. Urry (2002) notes how tourism experiences are shaped by the intersections of class, gender and ethnicity, and after exemplifying this with Chinese restaurants clustered in a small area of Manchester, concludes (2002: 9) that 'ethnic groups are important in the British tourist industry... and in some respects play a key role .... In recent years certain ethnic groups have come to be constructed as part of the "attraction" or "theme" of some places'.

These spatial links between tourism and ethnic exotica are not new, of course, nor are they confined to the current periods of globalization. Judd (2003), for example, has described the construction of tourist enclaves that accompanied the middle-class Grand Tour to Paris, Geneva, Rome, Florence, Venice and Naples and the emergence of the potential for mass tourism in Europe following the establishment of Thomas Cook in the mid-nineteenth century. On the other side of the Atlantic, Lin (1998: 174–6) has recounted how middle-class New Yorkers in the 1880s liked 'to go slumming in Chinatown', riding in 'rubbernecker vehicles' (also known as 'gape wagons'), with the term *rubbernecker* (for a gawking tourist) entering into American parlance during this era. Urban tourism has thus been linked to cultural diversity from the earliest days.

Nonetheless, globalization today is accompanied by a growing international movement of permanent and temporary immigrants (Castles and Miller 2003) and a corresponding explosion in the ethnic and cultural diversity of most contemporary Western cities. As Levitt (2001: 203) has put it, 'more and more residents of contemporary nation states are non-nationals or hold dual citizenship'. At the same time – and often building on centuries-old patterns of immigration – national and international urban tourism now is increasingly linked to the commodification of culture and ethnicity. There is a link between the two. Focusing on the increasing immigrant diversity, Hoffman (2003: 97) has argued that 'the pursuit of ethnic branding reflects the fact that minorities are the fastest growing (new) consumer population', while Judd (2003: 32) suggests that 'tourism overlaps with – indeed, is a product of – a globalized culture of consumption sustained by highly mobile workers and consumers'.

The major symbolic representation of the urban ethnic precinct is ethnicity and ethnic diversity. Yet what constitutes such an 'authentic' ethnic tourism experience within the city? What symbols are appropriate, who decides, and how? There is a fundamental contradiction here between the outdated ethnocultural stereotypes and tourist iconography in countries of immigration that usually depict a static homogeneity of immigrant or ethnic experience and the dynamic diversity of contemporary life in countries of immigrant origin. As Fainstein *et al.* (2003: 246) put it, 'the tension between differentiation and homogeneity makes for a contradiction and conflict in urban tourism regimes'. The remainder of this chapter explores the contradictions and conflicts that emerge from the creation of three such ethnic precincts in Sydney.

Ethnic precincts in Sydney

Sydney is Australia's most cosmopolitan city, with some 60 relatively large ethnic communities and more than 100 micro-ethnic communities (Collins and Castillo 1998). Table 4.1 shows the birthplace groups of first-generation immigrants whose populations in Sydney exceeded 20,000 as at the 2001 National Census. It shows that the largest immigrant groups, after those born in the UK, are from China, New Zealand, Vietnam, Lebanon, Italy, Hong Kong, India, Greece, Korea, Fiji and South Africa. In addition to these groups, there are another 170 immigrant communities in Sydney with populations of less than 20,000. Clearly ethnic diversity is a vital feature of contemporary Sydney.

Analysis of the (changing) spatial patterns of immigrant settlement in Sydney gives a clue to the physical location of the ethnic economy and the location of ethnic precincts. Most immigrant minorities – that is, those from non-English-speaking backgrounds, or 'NESB immigrants' – live in Sydney's southwestern suburbs (Collins and Poynting 2000). All of the local government areas (LGAs) with relatively high proportions of first- and second-generation NESB immigrants are located in those suburbs. However, even the LGAs with the highest immigrant concentrations are themselves very diverse and multicultural, with no one immigrant birthplace group dominating the population of any Sydney LGA. Over time,

Table 4.1 Sydney's population by birthplace*

Birthplace	Numbers
Australia	2,454,424
United Kingdom	183,991
China	82,029
New Zealand	81,963
Vietnam	61,423
Lebanon	52,008
Italy	48,900
Hong Kong	36,039
India	34,503
Greece	33,688
Korea	26,928
Fiji	25,368
South Africa	25,190

Source: 2001 Census
* Populations in excess of 20,000

immigrants tend to move house, and a pattern of 'ethnic succession' occurs. In the first three post-World War II decades, newly arrived minority immigrant groups often settled in poorer areas, replacing immigrants who had built up enough resources to move to more preferable locations and neighbourhoods (Burnley 1986). More recently, the shift of Australian immigration policy to the highly skilled and qualified has added a new process of ethnic succession in middle-class suburbs (Burnley 2001).

There are many ethnic places and spaces in Sydney today. Like so many Western cities with a minority immigrant history (Anderson 1995; Fong 1994; Lin 1998; Zhou 1992), Sydney has a prominent and long-established Chinatown in the downtown area, although most of Sydney's other ethnic precincts are located in the suburbs of southwestern Sydney. Sydney's ethnic precincts include Leichhardt (Little Italy), Campsie (Little Korea), Petersham (Portuguese) and Marrickville (once Greek, now Vietnamese) in Sydney's inner-southwestern suburban ring. In the middle-southwestern suburban ring, ethnic precincts include Auburn (Turkish quarter), Lakemba and Punchbowl ('Middle Eastern') and Bankstown (Asian and Middle Eastern). Cabramatta, in the Fairfield municipality, is even further from the city centre and has become an Asiatown. One exception to the location of Sydney's ethnic precincts in southwestern Sydney is the North Shore Chinese precinct of Chatswood, the centre of professional and well educated middle-class Chinese immigrants. In addition, the Bondi Beach area in Sydney's eastern suburbs has a prominent Jewish history and presence. Some of these areas, like many other suburbs across the breadth of Sydney, are tangible manifestations of ethnic diversity in public space, at least in terms of restaurants.[1] Some take the title 'precincts', others 'quarters', while others get no nomenclature at all. For the sake of brevity, only the ethnic precincts of Chinatown, Little Italy and Cabramatta will be explored in any detail in this chapter.

It is interesting to note here that the 'ethnic precinct' is related to immigrant minorities, not to the majority immigrant groups, which in Sydney's case are Anglo-Celtic. This stems in part from the fact that 'ethnicity' is (falsely) constructed in Australian discourses as something exclusive to the cultures of NESB or non-white immigrants (Castles *et al.* 1988; Hage 1998). This is partly because the familiar or the mainstream is not sufficiently exotic, partly because the patterns of settlement of the Anglo-Celtic majority are much less concentrated than those of minority immigrants, and partly because of the relatively low rate of entrepreneurship of the British and Irish majority as compared to many immigrant minorities in Sydney (Collins 2003). And, developing on from this, it may also be due to the lack of English, Irish, Scottish and Welsh restaurants.

Sydney has had a Chinatown since the 1860s (Anderson 1990) and, as in Chinatowns the world over (Zhou 1992), the major Chinese businesses were restaurants, grocery stores, market gardening, furniture and cabinet-making, and import/export. In the 1890s Sydney's Chinatown moved to the Gipps Ward, west of the central business district, and in the 1940s it moved to Campbell and Dixon Streets in the city, where it is still located today (Collins and Castillo 1998; Fitzgerald 1997). Immediately after the Second World War, Chinese immigrants continued their earlier presence in the vegetable and fruit retailing business and in running cafés and restaurants in metropolitan and rural areas across the nation (Collins 2002). Today Sydney's Chinatown is a vibrant and lively precinct.

Leichhardt has been the original home of Sydney's Italian immigrant community since the end of the nineteenth century. In 1885 the fishmonger Angelo Pomabello and the Bongiorno Brothers were among the first Italians to settle in Leichhardt. They opened a fruit shop on Parramatta Road. But it was not until the 1920s that a Little Italy began taking shape in the Leichhardt community. The Italian community expanded dramatically in the following years, and was further reinforced with a massive wave of migrants in the late 1950s and early 1960s. For the postwar Italian immigrants, Leichhardt offered cheap housing, proximity to employers of unskilled labour, Italian shops and other businesses. Religion and commerce were at the centre of this flourishing community. The Saint Fiacre church and parish, run by the Italian-speaking Capuchin Fathers, became the hub of Italian life in the area. As early as 1962 there were four Italian cafés in Leichhardt, and soon they were joined by other businesses such as fruit vendors, real estate agents, grocers, restaurants, hairdressers, bookmakers, butchers, pharmacies, shops, bakeries, jewellers, music shops and night clubs.

Since then, Italian-born entrepreneurs have reinforced Little Italy by adding to the cafés and restaurants along the strip. Burnley (2001) lists 325 Italian owned businesses in Leichhardt and neighbouring Five Dock. Of these, 72 were professionals (doctors, accountants, dentists, optometrists, solicitors and paramedics), 58 were light industrials (including terrazzo tiles and pasta food manufacture), while 190 were involved in general retail (including 33 restaurants, 18 cafés, 13 butchers and 11 pasticcerie).

The ethnic precinct in Cabramatta (dubbed 'Vietnamatta' by the sensationalist tabloid media) emerged after an inflow of Indochinese refugees from nearby migrant hostels (Collins 1991). Along John Street, which runs along the western side of Cabramatta railway station, a vibrant ethnic precinct has grown up with over 820 ethnic businesses and institutions. Ian Burnley (2001: 252) gives a vivid description of the many ethnic businesses featuring wide ranges of goods and services, including professional services, in 1988:

> bakeries, butcheries (at least 20), cake shops, children's clothiers, confectioneries, arts and crafts, dress materials and fabrics, bridal wear shops, adult clothing retailers and manufactures, electrical goods suppliers, fish markets (6), general food stores, take-away foods (10 shops),

fruit shops (12), many groceries, hair and beauty salons (10), herbalists (15), jewellers, laundries, newspaper proprietors, newspaper publishers, delicatessens and food importers and manufacturers. There were 30 medical practitioners, 15 dentists, several physiotherapists, over 20 accountants, several land agents, and more recently the growth of travel agencies as it became possible for Vietnamese and Chinese to revisit South-east Asia.

(Burnley 2001)

The owners of these businesses were Vietnamese (particularly ethnic-Chinese Vietnamese), as well as other Chinese, Laotians, Cambodians and residual Italians, Croats and Serbs.

Ethnic precincts in Sydney mainly develop their contemporary ethnic identity through the ethnicity of the entrepreneurs who own the businesses in these suburbs, and not primarily because of the area's current population composition. Relatively few Chinese immigrants live in Chinatown today, although this trend has been reversed by the spate of residential development in Sydney's downtown prior to the Sydney 2000 Olympics. By 2001 there were only 2000 people (out of a total Leichhardt population of 60,000) who were born in Italy, and two-thirds of those living in the municipality were born in Australia, many to immigrant parents. Indeed, more New Zealand-born people live in Leichhardt today than do Italian-born. Similarly, in Cabramatta the majority of businesses along John Street are owned by Vietnamese immigrants (often ethnically Chinese), even though the ethnic Chinese population in Cabramatta is less than 15 per cent (Burnley 2001). In all cases, it is the immigrant entrepreneurs who give the precincts their ethnic or immigrant identity. But these entrepreneurs could not exist without adequate co-ethnic and/or mainstream custom, nor could the precinct emerge without approval and support from relevant authorities.

The problem of credibility and authenticity

In Sydney, 'ethnic precincts' are urban or suburban high street agglomerations of ethnic enterprises, clustered together in a space which formally or informally adopts the symbolism, style and iconography of that ethnic group in its public spaces. Ethnic eating places – restaurants, cafés, gelaterie – and shopping establishments are the prime attraction to many precinct visitors, although ethnic professionals often occupy the upper floor spaces, providing services to co-ethnics. As Selby (2004: 28) reminds us, 'ethnic tourism enclaves' are presenting in many countries: 'The potential for "ethnic tourist attractions" is now widely recognised by local governments and tourist boards, encouraging the development of areas such as Brick Lane in London, Little Italy in Boston and the Balti Quarter in Birmingham'. Important questions arise. How do ethnic restaurants and other immigrant enterprises symbolize their ethnicity in the restaurant/shop décors, menus, signage and divisions of labour? How does a precinct as a whole develop as an 'ethnic' place/space? And what role do regulation and regulators play in developing the ethnic precinct, in shaping the symbolic representation of ethnicity within the precinct, and in place-marketing the precinct?

The raison d'être of developing an ethnic precinct is to create a place and a space with an appealing ethnic smell, sound and sensibility, such that visiting this place in the city means experiencing the culture of one of Sydney's minority immigrant communities. As Boyle (2004: 3) argues, 'The point is that many of us – and an increasing proportion of the Western world – want to experience it ourselves. We want it real'. The challenge of the ethnic precinct is to recreate/represent an *authentic* ethnic experience of place by developing a symbolism, an ambiance and an experience that *credibly* coincides with wider expectations about or images of that ethnic group. For an ethnic precinct to be a vital part of urban tourism, it must have credibility in the eyes of those locals and tourists who use the space. That credibility – which could be measured quantitatively by return visitation, by the economic success of the ethnic enterprises or by attendances at precinct ethnic festivals – is in turn linked to the extent to which the ethnic precinct is seen as authentic by locals and tourists.

The main social groups interacting in the ethnic precinct are the immigrant entrepreneurs, the customers (locals and tourists), the workers in ethnic enterprises, the other resident communities in the area, and the regulators and institutional actors in the urban planning and marketing processes. The problem with the concept of authenticity when applied to the ethnic economy is that it is subjective. Views about what is authentic in the precinct might differ. Thus, what constitutes an 'authentic' ethnic or cultural eating or tourist experience could vary according to the different standpoints of those who participate in the daily life of the ethnic precinct. We should not be surprised by this. As Meethan (2001: 27) has put it, symbols 'are multivocal, that is, they have the capacity to carry a range of different, if not ambiguous and contradictory meanings'.

Responding to the subjective, and hence complex, character of authenticity when applied to the tourist experience, Meethan (2001) has developed an authenticity matrix to differentiate between various kinds of tourists (existential, experiential, recreational and diversionary) and various forms of tourism (alienation, authenticity and holiday). Similarly, Wang (1999) distinguishes three forms of authenticity: *objective* (historical record), *constructive* (authenticity as negotiation and ascribed meaning within power relations) and *existential* (experiences at an individual level). Central to this approach is the view that tourists are not passive dupes, but active agents shaping their tourist experience, though within constraints. Selwyn (1996) distinguishes between the more objective-world 'cool authenticity' and the more experiential 'hot authenticity' of place consumers. Meethan (2001) reminds us that authenticity is a matter of negotiation and ascribed meaning. Authenticity is also a contradictory concept. As Boyle (2004: 4) has put it,

> the dominant cultural force of the century ahead won't just be global and virtual, but a powerful interweaving of opposites – globalization and localization, virtual and real, with an advanced guard constantly undermining what is packaged and drawing much of society behind them.

The advanced guards of authenticity are akin to Zukin's (1995) *critical infrastructure*.

One of the critical parts of an ethnic precinct is its outer façade. What constitutes an authentic Chinese/Italian/Vietnamese place and how do you develop it? Bryman (2004: 52) refers to the centrality of *theming* in contemporary consumption places and the contradictions inherent in such theming attempts:

> Critics of theming often disapprove of the use of symbols of nostalgia for thematic cues. Drawing on faux designs and histories, theming in terms of nostalgic references is often depicted as presenting a sanitized history, one that removes any reference to hardship and conflict in the cause of consumption.

However, while some will view the Disneyization of ethnic culture (Bryman 2004) in places such as Orlando's Epcot Center and global restaurant chains as plastic and inauthentic, others will be satisfied that the Disney representation of cultural diversity corresponds with their expectations and imaginations. In other words, ethnic theming may result in a Disneylandic, un-authentic precinct but, paradoxically, such an artificial image may substantially increase the credibility if visitors *expect* a Disneylandic precinct.

Each of Sydney's ethnic precincts has undergone a series of *ethnic facelifts* over time in an attempt to give it an authentic ethnic feel or theme. For Chinatown, the redevelopment of Dixon Street began in 1972 with the introduction of porticoes and lanterns, and even trash bins with 'traditional' Chinese symbols, to make the area more 'Chinese'. The precinct was to appear more consistent with the architectural motifs and symbols of ancient China (Anderson 1990). In the 1980s, Dixon Street was developed further as a pedestrian thoroughfare, Chinese dragons were erected at the Paddy's Market end of Dixon Street, and Chinese trees were planted along the streetscape. It was linked via the Chinese Gardens to the new Darling Harbour development (Fitzgerald 1997). Hong Kong Chinese capital financed much of this revamping.

Local governmental councils were also critical players in the development of Little Italy in Leichhardt, providing funds for 'facial makeovers' that included Italian signage and wider footpaths to produce the feel of the 'al fresco' (outdoor) Italian eating experience. Italian entrepreneurs were encouraged to redevelop their restaurants and businesses to take on Tuscan façades and upstairs balconies, supplementing the 'tables on the footpaths' at ground level. Leichhardt Council also approved and promoted the building of the Leichhardt Italian Forum – a large residential and commercial development that recreated the Italian 'village feel', complete with four floors of residences with Juliet balconies, overlooking and encircling a large piazza where the tables of Italian restaurants had room to spread out under the stars. The Forum even featured clock towers, wandering Italian musicians and a central fountain.

In Cabramatta, as in Chinatown, attempts have been made by local and state policymakers to redevelop the shopping precinct to attract more customers and visitors from outside the area. In the early 1980s, the Cabramatta Chamber of Commerce – which at that time had no Vietnamese entrepreneurs on it – received a sizeable grant from Fairfield City Council to develop a plaza area along John Street. Then, in the late 1980s, a campaign was instigated, with a brief to 'change unfavourable images, to promote the acceptance of the Indo-Chinese community and foster multicultural activities such as the Fan Festival, the Dragon Boat Race, an international cabaret and "good eating"' (Burnley 2001: 248). More recently, government authorities have intervened with a crime and safety policy for the Cabramatta precinct, a matter explored on page 79.

Another aspect of the credibility of ethnic precincts relates to the experience of the tourist in the immigrant enterprises – and particularly the restaurants and cafés – that dominate the streetscape of every ethnic precinct. Credibility is thus also linked to the ethnicity of the entrepreneurs (mostly Italian in Little Italy, Chinese in Chinatown and Chinese or Vietnamese in Cabramatta) and how they stage their ethnicity, for example through the real or alleged 'authenticity' of the ethnic façades of individual restaurants, cafés and shops. Bryman (2004) argues that ethnically themed restaurant chains such as the Olive Garden restaurants (Italian) or Ricardo's Mexican restaurants 'involve playing up and tweaking symbols of ethnicity and in particular food-related expression'. For some customers, credibility may also be drawn from the ethnicity of the people who own and work (cook, serve) in the immigrant enterprise and from the authenticity of the products (menus, cooking ingredients) and goods delivered. Clearly issues of authenticity and credibility are subjective, mutually contradictory, renegotiated and redefined over time.

The problem of legitimation

In the previous section, the authenticity and credibility of the ethnic precinct was explored mainly from the point of view of the tourist (local, national or international). This section explores a related issue: the credibility of the ethnic precinct in the eyes of the co-ethnic community itself – that is, Chinese views about Chinatown, Italian views about Little Italy – and in the different eyes of key stakeholders, regulators and local authorities. In other words, how *legitimate* are ethnic precincts – having been designed to reproduce some authentic ethnic experience of place in the eyes of the host society and its visitors – to immigrants of a similar cultural background who often have pre-migration experience of the authentic? And how do policy-makers decide about the legitimacy of icons and symbolizations of ethnicity? This issue is particularly important given that the development of precincts involves local and provincial government authorities that negotiate with community and business representatives about plans to rejuvenate the precincts. But which representatives are 'authorized' to claim this ethnic authenticity?

This is not a trivial issue. For example, there are more than 100 ethnic Chinese community organizations in Sydney alone. What voice is representative, is authentic? How is this decided? This reality of complex, diverse and changing Chinese *communities* in Sydney flies in the face of simplistic attempts to construct *the* authentic Sydney Chinatown. State and local government attempts to redevelop Sydney's Chinatown have prompted internal struggles within the Chinese community over the right to gain representation on the relevant development and planning committees (Anderson 1990). More generally, Australian multiculturalism has established a hierarchy of 'ethnic leaders' who are consulted by governments and tend to favour certain, often conservative, ethnic community organizations and community leaders over others (Jakubowicz 1984). The reality of Chinese immigrant communities the world over is that they are complex, diverse and heterogeneous. This does not sit well with the ideal of ethnic marketing, which *badges* a Chinese precinct with a simple common motif. The same issues can be raised for Little Italy and Cabramatta.

The development of Sydney's ethnic precincts of Chinatown, Little Italy and Cabramatta has largely been shaped by local government authorities, thus yielding interesting case studies of the ways in which regulation shapes ethnic entrepreneurial (Kloosterman and Rath 2001) and urban tourism developments (Hoffman *et al.* 2003).

Anderson (1990: 150) argues that Sydney's Chinatown has been revitalized in ways that reflect white Australia's image of *Chinese-ness*: 'Making the area more "Chinese" …[meant] making the area appear more consistent with the architectural motifs and symbols of ancient China'. In other words, attempts by city councils in Sydney to 'create' a Chinatown in an image that would attract tourists have resulted in façades, monuments and facelifts reflecting stereotypical images of a homogeneous 'Chinese-ness' that exists only in the 'white gaze'. This is an argument made, too, about Chinatowns in other places like New York (Lin 1998: 173): 'In the process of retrofitting Chinatown for popular consumption, these outsiders deliberately manipulated reality to suit the imaginary expectations of

Western observers'. During this process of 'developing' and constructing Chinatown, the Chinese were viewed as a homogeneous Other, rather than as a community like all others – divided along regional, class and commercial lines.

This problem of legitimation impacts on the extent to which co-ethnics frequent the ethnic precinct. To some people, the authenticity of the ethnic precinct and its ethnic restaurants is reflected in who goes to the precinct and to its shops and restaurants. The presence of large numbers of co-ethnic locals and visitors to the ethnic precinct (Italians to Little Italy, Indochinese to Cabramatta, Chinese to Chinatown) are important to tourists, because they give the streetscape a more authentic feel. The restaurant where lots of co-ethnics are eating is likely to be experienced by unfamiliar tourists as a supremely authentic restaurant, just as the precinct as a whole gains authenticity from the scale of co-ethnic custom. Little Italy acquires much of its Italian feel from the passing parade of Italian families shopping, eating in its restaurants, visiting Italian professionals or just walking arm-in-arm along Norton Street with gelato in hand. When Italy wins or loses World Cup soccer matches, Little Italy is where you will find happy or sad Italian supporters. Authenticity is further boosted by the role of the ethnic precinct as the place where ethnic festivals are held. Chinatown and Cabramatta both hold a series of festivals to mark events on the Chinese lunar calendar, while Cabramatta hosts many other ethnic community events, national days and other public celebrations of ethnic diversity. Little Italy holds an annual Norton Street Festival each March or April. Norton Street is closed and lavishly decorated in the red, white and green Italian colours. In place of cars, food and market stalls, art exhibitions and other entertainments regularly attract over 100,000 people, highlighting the popularity of this event (Collins and Castillo 1998).

In current fieldwork on four ethnic precincts in Sydney, Kunz (2004) has interviewed precinct customers as well as ethnic entrepreneurs and regulators. His random sample of tourists in the precincts was stratified into three distinct groups: co-ethnics (Chinese in Chinatown, Italians in Little Italy, Turkish in Auburn), co-cultural non-co-ethnics (non-Turkish Muslims in Auburn; 'Asians' in Chinatown) and others (the rest of mainstream cosmopolitan society). Kunz has found that views about what constitutes ethnic authenticity vary considerably between these three categories of customers in the ethnic precinct. He has also found evidence of both conflict and cooperation between the immigrant entrepreneurs in the precinct, as the competitive realities of business sometimes override ethnic solidarity.

The problem of control

The issue of tourist safety is central to any government tourist strategy. No one wants to go to a place where their or their family's safety is put at risk.

Control and surveillance therefore play an integral part in the development of tourism in general, especially in potential tourist precincts such as ethnic neighbourhoods. Body-Gendrot (2003: 39) has emphasized the importance of 'techniques of social control and security' required by mega-event tourism like the Olympic Games or World Cup soccer, while Judd (2003) observes that building tourist places as fortress spaces is one response to managing issues of tourist safety. Borrowing from Foucault, Edensor (1998, 2001) notes that in shopping malls there is a 'remorseless surveillance through panopticon visual monitoring'. Shopping is encouraged, but, as Judd (2003: 29) points out, 'aimless loitering is discouraged or forbidden'.

A number of aspects of control and surveillance are relevant to Sydney's ethnic precincts. The first involves the historical construction of minority immigrant communities as the criminal Other and a threat to the safety of the host society (Collins *et al.* 2000; Poynting *et al.* 2004), so that the places and spaces where they concentrate also get labelled unsafe and criminal.

Ironically, this criminal feel can also be an attraction to tourists. Chinatowns the world over have always had a criminal aspect. In New York's Chinatown, Chinese tongs, or gangs, are involved in crime and control the streets, and gambling is still a popular leisure activity in the Chinese community. According to Kinkead (1993: 47), tourists in the early twentieth century

> ...went to Chinatown to ogle vice: guidebooks warned of the immorality and filth of the quarters. The sightseers hired guides to show them opium dens, slave girls, and sites of lurid tong murders. Bohemians visited to smoke opium and drift away into hazy dreams.
>
> (Kinkead 1993: 47)

The focus on the ethnic-precinct-as-crime-precinct in Sydney today, however, is not on Chinatown, but on Cabramatta, which has developed a reputation as the heroin capital of Sydney. In late 2001, Sydney's tabloid newspaper the *Daily Telegraph* ran a series of articles about Cabramatta's drug problems under the following headline: 'Cabramatta was favoured by tourists until the drug lords moved in. Now locals say they are winning it back'. Here is one news item: 'Five men, a 21-year-old woman and an eight-year-old boy were sprayed by 10 of 15 bullets fired from a semi-automatic handgun during the attack, at Cabramatta's New World Seafood Restaurant on May 25'.[2]

Such incidents pose a problem of safety for visitors, and create an image in sharp contradistinction to the one promoted by local government authorities. A glossy brochure targeting visitors to the city claims that 'Cabramatta is a day trip to Asia ... Here, an hour from the centre of Sydney, is an explosion of Asian colour – a bustling marketplace offering all the ingredients for a banquet for the senses'. Local expert guides accompany visitors on a walk through Cabramatta, helping build an appreciation for the various types of Asian products sold there. More recently the Fairfield City Council launched a CD-guided driving tour of the ethnic sites and features of Cabramatta. The results are impressive if we are to believe the Council: more than 350,000 visitors from Australia and overseas visit Cabramatta every year, spending more than €83 million in local shops and on services. Representatives from the local government even claim that for every 17 international visitors, one extra job is created, making tourism a major employer.

Jacobsen sharply captures the contradictions of Cabramatta:

> It is the nation's drug capital, and a crime hot spot. But Cabramatta is also one of Australia's most culturally diverse places and this morning it is showing off its spiritual side with the launch of a driving tour of the Fairfield region. The tour includes Vietnamese, Chinese, Cambodian and Laotian temples; Croatian, Assyrian, Russian and Serbian orthodox churches; a Dutch shop and a Turkish mosque. It will be launched by the Premier, who only weeks ago upset ethnic groups with comments about gangs and immigration, and comes as Middle East migrants are facing increased hostility after the US terrorist attacks.[3]

Many of the problems of ethnic precincts in Sydney relate not to ethnicity per se, but to socioeconomic status. This stems from their predominant location in the working-class southwestern part of the city, a region with the highest unemployment and poverty rates in Sydney, and the one suffering most from the way that globalization is decimating Australia's manufacturing industry (Collins and Poynting 2000). In 2001, some 150 homeless people were sleeping rough on Cabramatta's streets, while over 100 youth were dependent on community support to survive.[4]

This contradiction between Cabramatta as vital Asian/Vietnamese ethnic tourist experience and as heroin capital is apparent. Crime and fear of crime is not good for business – ethnic or otherwise. Government has responded with public surveillance cameras and a policing strategy that has greatly increased the number of officers on foot patrol. Reports suggest that crime rates have fallen in the area. Sofios further noted that 'a few restaurants have started opening later as interest increases in the suburb's wide array of Asian foods, refusing to close at 6 pm and give in to the shadowy figures that still roam the streets at night'.

Surveillance in ethnic precincts also relates to the workplace: the ethnic enterprise. Bryman (2004) has stressed that control and surveillance (of both consumers and workers) are central to the Disneyization of the whole of society. He cites Susan Willis (1995), who claimed that 'there is probably more surveillance per square inch (both technological and human) [in Disney World] than in any of today's underfunded public prisons'. Disney uses employees disguised as tourists ('shoppers' who 'shop' the employees – that is, report on them), and workers also face surveillance through various technological means and through customer feedback.

Of course, ethnic precincts in Sydney and elsewhere are mainly sites of small business enterprises rather than global corporations such as Disney, but the issue of controlling workers is also a critical one for small immigrant entrepreneurs. Many employ family or co-ethnic labour for reasons of trust. This also helps give the ethnic enterprise authenticity, reinforcing its ethnic feel. Another labour issue in ethnic enterprises is cost. Many ethnic restaurants and other enterprises are economically marginal (Collins *et al.* 1995) and hence are often the site of exploitative labour relations, such as unpaid family labour or undocumented co-ethnic labour. There are many instances of ethnic restaurants in Sydney employing undocumented immigrants to wash dishes, assist in preparing and serving meals, and perform cleaning.

Conclusion

The ethnic precinct is one key site of the immigration/ethnic diversity/tourism interface in the city. This chapter has examined three of Sydney's ethnic precincts to begin to unravel the social, economic, political and cultural dynamics which inform their historical development and their contemporary prospects for tourism. One important finding is that immigrant entrepreneurs are central to the creation of ethnic precincts in the first instance, as well as to their vitality and authenticity in the longer term. Tourism therefore needs to delve more into the literature on ethnic entrepreneurship if it is to fully understand the tourist dynamics of commodified ethnic spaces and places in the city. Another key finding here is that local and provincial governments and regulators play critical roles in developing and marketing ethnic precincts. This is consistent with the regulation theory of tourism (Hoffman *et al.* 2003), currently the most rigorous framework for understanding contemporary urban tourism – though it, too, needs to focus more attention on the regulatory issues impacting on immigrant entrepreneurs and enterprises (Kloosterman and Rath 2001).

More specifically, this chapter has explored the contradictions that underlie the development of Sydney's ethnic precincts and their marketing to locals and tourists alike. Three problem areas were identified, each with its own set of contradictions. First, there is the problem of credibility and authenticity, which involves who is 'authorized' to claim authenticity, how that authenticity is symbolized and what employees and employers in ethnic enterprises have to do to generate that authenticity. Because authenticity is subjective, more research is needed in Sydney to investigate the customer or 'demand' side of the commodification of ethnic diversity, including new research on the characteristics of consumers in ethnic precincts that is sufficiently disaggregated to allow distinctions between local, city, national and international tourists, the type of tourism, and the ethnic background and immigrant status of the tourists themselves (if appropriate). Of interest here are the ways and means through which different people gain knowledge of and form expectations about different ethnicities in general, and Sydney's ethnic precincts in particular.

This leads to the second contradiction, arising from the problem of how legitimate a precinct can be in the eyes of the co-ethnic community, other locals and tourists if it has been developed by deliberate regulation, planning and government intervention. Is it possible to develop ethnic precincts such as Sydney's Chinatown without necessarily reproducing white stereotypes of the ethnic Other?

Third, there is the problem of control and the ways that crime in ethnic precincts threatens the safety of the ethnic tourist experience. Control also relates to businesses, and involves issues of undocumented or exploitative (often co-ethnic) labour that stem from the blurred boundaries between the formal and informal economy in ethnic precincts and immigrant enterprises.

If a relationship exists between immigration, ethnic diversity and tourism in the city, it will be most obvious in, though not confined to, ethnic precincts. We still do not know enough about the nature or extent of tourism to Sydney's ethnic precincts, nor about the effectiveness of attempts to promote such tourism. In view of the centrality of ethnic food to ethnic precincts, we also need comparative studies of ethnic precincts and ethnic restaurants in different cities and different countries, to compare how ethnicity is staged and how tourists respond. Other comparative studies are needed of the ethnic tourism infrastructure – precincts, landmarks, churches, mosques, ethnic organizations, ethnic shopping and ethnic heritage – and of the ways in which popular culture constructs images of ethnicity through tourist guides, recipes, critical restaurant reviews in the media and the like. Finally, we need to investigate more closely the direct and indirect links between tourism and immigration.

Notes

1 By the mid-1980s, Chinese cafés or restaurants were a feature of the Australian landscape in suburbia and country towns (Collins *et al.* 1995). According to Chin (1988), there were 700 Chinese-operated cafés in New South Wales, with 300 in Sydney at that time, most employing Chinese labour.
2 J. Kidman, 'Gunman "fled overseas"', *Sydney Sun Herald*, 30 June 2002.
3 G. Jacobsen, 'Spiritual tour aims at toleration in ethnic heartland', *Sydney Morning Herald*, 17 September 2001.
4 S. Sofios, 'Suburb winning the war on drugs', *Daily Telegraph* (Sydney), 2 October 2001.

References

Anderson, K.J. (1990) 'Chinatown re-oriented: a critical analysis of recent redevelopment schemes in a Melbourne and Sydney enclave', *Australian Geographical Studies*, 28: 131–54.

Anderson, K.J. (1995) *Vancouver's Chinatown: racial discourse in Canada, 1875–1980*, Montreal: McGill–Queens University Press.

Body-Gendrot, S. (2003) 'Cities, security, and visitors: managing mega-events in France', in L.M. Hoffman, S.S. Fainstein and D.R. Judd (eds) *Cities and Visitors: regulating people, markets, and city space*, Oxford: Blackwell.

Boyle, D. (2004) *Authenticity: brands, fakes, spin and the lust for real life*, London: Harper Perennial.

Bryman, A. (2004) *The Disneyization of Society*, London: Sage.

Burnley, I. (1986) 'Immigration: the post-war transformation of Sydney and Melbourne', in J. Davidson (ed.) *The Sydney-Melbourne Book*, Sydney: Allen & Unwin.

Burnley, I. (2001) *The Impact of Immigration on Australia: a demographic approach*, South Melbourne: Oxford University Press.

Castles, S. and Miller, M.J. (2003) *The Age of Migration: international population movements in the modern world*, 3rd edn, New York: Guilford Press.

Castles, S., Kalantzis, M., Cope, B. and Morrissey, M. (1988) *Mistaken Identity: multiculturalism and the demise of nationalism in Australia*, Sydney and London: Pluto Press.

Chin, K.H. (1988) 'Chinese in modern Australia', in J. Jupp (ed.) *The Australian People: an encyclopedia of the nation, its people and their origins*, Sydney: Angus & Robertson.

Collins, J. (1991) *Migrant Hands in a Distant Land: Australia's post-war immigration*, Sydney and London: Pluto Press.

Collins, J. (2002) 'Chinese entrepreneurs: the Chinese diaspora in Australia', *International Journal of Entrepreneurial Behaviour and Research* 8(1/2): 113–33.

Collins, J. (2003) 'Australia: cosmopolitan capitalists down under', in R. Kloosterman and J. Rath (eds) *Immigrant Entrepreneurs: venturing abroad in the age of globalization*, Oxford and New York: Berg.

Collins, J. and Castillo, A. (1998) *Cosmopolitan Sydney: explore the world in one city*, Sydney: Pluto Press.

Collins, J. and Poynting, S. (eds) (2000) *The Other Sydney: communities, identities and inequalities in Western Sydney*, Melbourne: Common Ground.

Collins, J., Gibson, K., Alcorso, C., Tait, D. and Castles, S. (1995) *A Shop Full of Dreams: ethnic small business in Australia*, Sydney and London: Pluto Press.

Collins, J., Noble, G., Poynting, S. and Tabar, P. (2000) *Kebabs, Kids, Cops and Crime: youth ethnicity and crime*, Sydney: Pluto Press.

Connell, J. (ed.) (2000) *Sydney: the emergence of a world city*, New York: Oxford University Press.

Costa, N. and Martinotti, G. (2003) 'Sociological theories of tourism and regulation theory', in L.M. Hoffman, S.S. Fainstein and D.R. Judd (eds) *Cities and Visitors: regulating people, markets, and city space*, Oxford: Blackwell.

Edensor, T. (1998) *Tourists at the Taj: performance and meaning as symbolic site*, London: Routledge.

Edensor, T. (2001) 'Performing tourism, staging tourism: (re)producing tourist space and practice', *Tourist Studies*, 1: 58–81.

Fainstein, S.S., Hoffman, L.M. and Judd, D.R. (2003) 'Making theoretical sense of tourism', in L.M. Hoffman, S.S Fainstein and D.R. Judd (eds) *Cities and Visitors: regulating people, markets, and city space*, Oxford: Blackwell.

Fitzgerald, S. (1997) *Red Tape, Gold Scissors*, Sydney: State Library of New South Wales Press.

Fong, T.P. (1994) *The First Suburban Chinatown. the remaking of Monterey Park, California*, Philadelphia: Temple University Press.

Gabaccia, D.R. (1998) *We Are What We Eat: ethnic food and the making of Americans*, Cambridge MA: Harvard University Press.

Hage, G. 1998. *White Nation: fantasies of White supremacy in a multicultural society*, Sydney: Pluto Press.

Halter, M. (2000) *Shopping for Identity: the marketing of ethnicity*, New York: Schocken.

Hoffman, L.M. (2003) 'Revalorizing the inner-city: tourism and regulation in Harlem', in L.M. Hoffman, S.S Fainstein and D.R. Judd (eds) *Cities and Visitors: regulating people, markets, and city space*, Oxford: Blackwell.

Hoffman, L.M., Fainstein, S.S. and Judd, D.R. (eds) (2003) *Cities and Visitors: regulating people, markets, and city space*, Oxford: Blackwell.

Jakubowicz, A. (1984) 'Ethnicity, multiculturalism and neo-conservatism', in G. Bottomley and M. de Lepervanche (eds) *Ethnicity, Class and Gender in Australia*, Sydney: George Allen & Unwin.

Judd, D.R. (2003) 'Visitors and the spatial ecology of the city', in L.M. Hoffman, S.S. Fainstein and D.R. Judd (eds) *Cities and Visitors: regulating people, markets, and city space*, Oxford: Blackwell.

Kearns, G. and Philo, C. (eds) (1993) *Selling Places: the city as cultural capital past and present*, Oxford: Pergamon.

Kinkead, G. (1993) *Chinatown: a portrait of a closed society*, New York: Harper Perennial.

Kloosterman, R. and Rath, J. (2001) 'Immigrant entrepreneurs in advanced economies: mixed embeddedness further explored', *Journal of Ethnic and Migration Studies*, 27(2): 189–202.

Kloosterman, R. and Rath, J. (eds) (2003) *Immigrant Entrepreneurs: venturing abroad in the age of globalization*, Oxford and New York: Berg.

Kunz, P. (2004) 'Fieldwork in four Sydney ethnic precincts', unpublished manuscript.

Lash, S. and Urry, J. (1994) *Economies of Signs and Space*, London: Sage.

Levitt, P. (2001) *The Transnational Villagers*, Berkeley: University of California Press.

Li, W. (1998) 'Anatomy of a new ethnic settlement: the Chinese ethnoburb', *Urban Studies*, 35(3): 479–501.

Light, I. and Gold, S.J. (2000) *Ethnic Economies*, San Diego: Academic Press.

Lin, J. (1998) *Reconstructing Chinatown: ethnic enclave, global change*, Minneapolis: University of Minnesota Press.

Meethan, K. (2001) *Tourism in Global Society: place, culture, consumption*, New York: Palgrave.

Poynting, S., Noble, G., Tabar, P. and Collins, J. (2004) *Bin Laden in the Suburbs: criminalizing the Arab Other*, Sydney: Federation Press.

Rath, J. (ed.) (2000) *Immigrant Business: the economic, political and social environment*, Basingstoke/New York: Macmillan/St Martin's Press.

Selby, M. (2004) *Understanding Urban Tourism: image, culture and experience*, New York: I.B. Tauris.

Selwyn, T. (ed.) (1996) *The Tourist Image: myths and myth making in tourism*, Chichester: Wiley.

Suvantola, J. (2002) *Tourist's Experience of Place*, Aldershot: Ashgate.

Urry, J. (2002) *The Tourist Gaze*, 2nd edn, London: Sage.

Waldinger, R., Aldrich, H., Ward, R. *et al.* (1990) *Ethnic Entrepreneurs: immigrant business in industrial societies*, Newbury Park, London and New Delhi: Sage.

Wang, N. (1999) 'Rethinking authenticity in tourism experience', *Annals of Tourism Research*, 26(2): 349–70.

Willis, S. (1995) *Inside the Mouse: work and play at Disney World*, Durham NC: Duke University Press.

Zhou, M. (1992) *Chinatown: the socioeconomic potential of an urban enclave*, Philadelphia: Temple University Press.

Zukin, S. (1995) *The Cultures of Cities*, Cambridge MA and Oxford: Blackwell.

Part II
Immigrant workers

5 Caterers of the consumed metropolis

Ethnicized tourism and entertainment labourscapes in Istanbul

Volkan Aytar[1]

We should thank Orientalism, really. It made us win the Eurovision Song Contest, and it makes our city a tourist attraction for the Westerners.[2]

This chapter analyses the process of social bifurcation and ethnicization of the labourscape in Istanbul's globalizing urban tourism and entertainment sector. Tourism in Istanbul is part of a larger amalgamation that includes tourism, entertainment and leisure settings in the city. This sector, which has risen to prominence in recent decades, has largely been shaped by the changeover in Turkish society to an increasingly diversified consumerism and the transformation of Istanbul into a city of urban consumption – a globalizing tourist and entertainment city catering to both tourists and residents. Before I discuss the role of ethnicization in the divided stratification of this labourscape, let me first describe the locus of these transformations and trace the major developments that have shaped the changes in Turkey and in Istanbul during the past quarter of a century.

Istanbul and Turkey since 1980: social and demographic change and urban transformation

Since the early 1980s, Turkey has undergone a major shift from a development model based on import-substituting industrialization, led by a protectionist state, to an export-promoting, market-oriented growth strategy (Arıcanlı and Rodrik 1990a, 1990b). Previous populist welfare policies that had led to rising real wages for some segments of society (Keyder 1987) were gradually shelved, and labour organizations were muted (Önder 1997). There was a fall in the number of public sector jobs, which had been a major source of employment. Industrial employment also shrank, especially for rural-to-urban migrants (Ayata 1997).

These transformations were superimposed on the persistent and growing pressures from demographic changes and population movements. Since the late 1940s at least, agricultural mechanization and market integration had been putting increasing strain on smaller-scale farming and had touched off a massive flow of rural-to-urban migration (Akşit 1998; Karpat 1976; Kıray 1998). This migration accelerated further after the 1960s. Coupled to a population growth rate amongst the highest in the world (Kıray 1998), the internal migration created conditions for rapid urbanization, mounting urban populations, and urban transformation. Istanbul was one of the cities receiving the bulk of the rural migration (Danielson and Keleş 1984). Up to the 1980s, the absorption of migrants into urban areas was less problematic than thereafter, as most of them found employment in the industrial and sometimes the public sectors, as well as in some parts of the service sector.

As a consequence of political and economic changes induced by the neoliberal reforms, the migrants arriving in urban areas, and in Istanbul in particular, during the 1980s and 1990s found themselves in an entirely different environment. In contrast with the earlier voluntary migrants who had left their provinces chiefly because of economic push and pull factors, many of the new migrants were internally displaced people, forced out of south-eastern Anatolia after 1984 by the protracted fighting between the Turkish security forces and the separatist rebels of the Kurdistan Workers' Party PKK (Çelik 2002; Kirişçi 1998; on other aspects of outmigration in the south-east, see Akşit *et al.* 1996). As rural migration and rapid urbanization continued during the 1980s and 1990s,[3] migrants were faced with deteriorating employment opportunities, especially in industry (Ayata 1997). In combination with the ongoing growth in population – especially the younger population – this translated into mounting unemployment problems, which magnified the effects of a worsening income distribution (Özmucur 1995; Şenses 1990).

Istanbul was transformed from a major base for large-scale manufacturing firms[4] into Turkey's globalizing centre for finance and banking (Keyder and Öncü 1994). Istanbul increased its hold over the Turkish economy and became the preferred location for multinational corporations trying to make inroads into the Turkish market (Pérouse 1999, 2000). As manufacturing investment sank (Aksoy 1996; Sönmez 1996), the number of branches of multinational companies shot up dramatically, and most firms providing producer services (consultancy, banking, insurance) chose Istanbul as their base of operations.[5] The overall share of the service sector increased substantially between 1980 and 2000. Confirming this transformation, Keyder (1999: 19) has observed that in the Istanbul of today 'there is a flourishing service sector in marketing, accounting and management, telecommunications, banking and finance, transport, insurance, computers and data processing, legal services, auditing, accounting, consulting, advertising, design and engineering'.

All these developments have implications for Istanbul's recent transformation into a 'globalizing city', one increasingly subject to global flows of capital and goods, as well as culture.[6] Furthering this process since 1980 has been a gradual decentralization of Turkish urban policy. This has enabled the Greater Municipality of Istanbul (co-operating at various times with the national government, international institutions like the World Bank, corporate and business groups, or other actors) to launch major schemes to transform the infrastructure and appearance of the city to make it more 'attractive' for global audiences, be they foreign investors or international tourists. Tourism development has been one of the key factors in globalizing the reach of Istanbul (Pérouse 2000).

The globalizing orientation has been mirrored in changing population configurations and in transformations of the urban space. The persistent widening of the income distribution in Istanbul has worsened social polarization. The most extreme distribution of income was reached in the mid-1990s, with the richest 20 per cent of the population controlling 64 per cent of the income and the poorest 20 per cent receiving a mere 4.2 per cent (Sönmez 1996). Istanbul's urban space became increasingly fragmented, with divisions along new fault lines. A new globalizing population stratum emerged (Aksoy 1996), as those employed in the globalizing sectors of the economy moved towards 'world-class incomes' (Kandiyoti 2002). The poorest strata, on the other hand, became further marginalized and were excluded from the process of articulation into the world economy.

A new setting of tourism and entertainment: consumption in a heterogeneous metropolis

As a further consequence of Istanbul's globalizing orientations and changing demographics, a new service sector consisting of tourism and entertainment establishments emerged. The alternative, 'informal' globalization of Istanbul as observed by Keyder and Öncü (1994) has not only had important repercussions on lifestyles, consumption patterns and urban space, but has also generated a mix-and-match heterogeneity and diversity. I would argue that this heterogeneity and diversity has largely shaped the new Turkish urban consumerism. This, in turn, has helped to dramatically 'upgrade' the entertainment sector and, along with it, the tourist spaces and amenities in Istanbul.

Urban tourism spaces and establishments catering to foreigners have flourished and diversified in Istanbul since the mid-1980s.[7] These should be situated in a more general framework of 'pleasure-oriented environments' in the increasingly consumption-oriented Istanbul setting. I borrow here the idea of placing tourism in this larger context, and I argue that tourism and other forms of leisure can be viewed on a continuum (Carr 2002). Swain has similarly argued that 'neither [tourism nor leisure] can be adequately understood without reference to the other' (in Carr 2002: 975).

Indeed, new entertainment spaces catering mainly to Istanbul's resident globalizing stratum have also begun to multiply and diversify. I am thinking here of numerous new bars, dance halls, clubs, cafés and similar venues.

Yenal (2003) has charted the proliferation of Istanbul's music scene over the past 15–20 years. He argues that in terms of such entertainment venues, Istanbul has become a 'polycentric' city. Following up on Aksoy's (1996) observation that the numbers of entrepreneurs and high-skilled high-paid white-collar professionals have multiplied in the globalized sectors of Istanbul's economy (business services, media, finance and tourism), I conclude that this globalizing stratum now constitutes the primary clientele of the new entertainment spaces.

Although one could claim that this stratum is numerically small, its importance far outweighs its size. Erkip (2000: 373–4) has argued that with the growth of the service sectors, the highly paid, well educated professional segments of the Istanbul population serve as a bridge between 'global and local values and lifestyles', and that cultural events in particular 'play their role in reflecting the distinctive taste of a global elite'. Yenal (2000: 5) has similarly pointed out that, however small in size, the new middle classes are the 'major receivers of global cultural flows ... [and] have trend-setting consumption patterns'.

Though set in a sociohistorical context quite different from the post-industrial turn as experienced in the West, Istanbul's upper and upper-middle classes and their class fractions are increasingly positioning themselves socially and culturally vis-à-vis others through the media of consumption and critical appreciation (Bourdieu, [1979] 1984). Scholars have convincingly argued that – akin to the developments in advanced Western economies – the culture of consumption has lately become instrumental in establishing and maintaining social boundaries in Turkey (for various discussions, see Kandiyoti 2002; Navaro-Yashin 2002; Öncü 1997; Yenal 2000).

In addition to being a heritage city and a cultural and historic site, Istanbul has thus evolved into a many-faceted city of urban consumption. One can now attend jazz festivals and independent film festivals; shop at world-renowned brand stores, alongside street bazaars; dine on anything from traditional Anatolian to Portuguese or from Chinese to vegan cuisine; listen and dance to samba, salsa, raï, reggae and new gay techno; and work out in upscale health and fitness clubs.

Although the growing significance of consumption in Turkey has been occasionally researched, studies that analyse this consumption within its sociohistorical, political, economic and cultural context are notably few. Most existing studies are either consumer- or provider-focused or they examine the social, cultural and symbolic significance of consumption, devoting little attention to the wider impacts that consumerism has on employment, demographics and policy. Nor has the boom in entertainment been well researched as an integral part of this consumerism, and almost no studies have been done on the changing nature of service sector employment as a whole, or on the nature of employment specifically in the entertainment and tourism business.

Tourism and entertainment employment the world over: helpful for understanding Istanbul's case?

Given that Istanbul continued to receive internal migration and began to experience growing immigration from abroad during a period when industrial and public sector jobs were dwindling, we need first to clarify the links between demography, migration and employment, learning from other case studies.[8] By identifying the 'human pool' of potential labour that tourism and entertainment businesses in other countries tend to employ, we might begin to explain the specific forms that employment in Istanbul's tourism and entertainment industry has acquired.

According to the literature, service sector employers throughout the world attract not only 'local native' workers, but also internal migrants and external immigrants. This workforce does not form a holistic, homogeneous entity, but a jumbled assortment of people with widely differing social, educational and skill backgrounds. As Li and colleagues (1998: 131–2) have shown, migrant workers in the service sector tend to be 'highly differentiated, involving, on the one hand, high-wage and highly skilled staff... [and] on the other hand, very large numbers of low-wage employees'. The former 'were drawn largely from developed countries... while [the] less skilled... were drawn from less developed neighboring labor markets'. Skill requirements in tourism vary widely; employees range from knowledge workers and managers to menial labourers with few or no qualifications.[9] This differentiation has far-reaching implications for migration patterns in tourism employment (Aitken and Hall 2000). Hall and Williams (2002) have argued that, owing to the conflation of production and consumption that is characteristic of tourism, tourism-related migration includes three types of people: unskilled labourers who are cheap to employ and are hired to provide routine services; managerial workers with specialist skills that may be scarce in the local labour pool; and entrepreneurs seeking to establish small- or medium-scale businesses, often catering to niche markets or lifestyle considerations.

The first set of workers may be recruited either from the local labour market or from the pool of migrant and immigrant labourers that are attracted to such jobs. The second category is more characteristically staffed by skilled local or immigrant employees, such as perpetually mobile Western 'expatriates'. The same may apply partly to the third category, especially as the niche markets or lifestyle considerations play a more determining role. Particularly in the development of the latter two categories, the emergence of new skilled labour markets which are less subject to national or centralized managerial control and regulation is an important factor (Iredale 2001). Highly skilled immigrants and expatriates follow a recent trend of increased permanent or long-term migration of professionals from developed to developing countries, where they may find more self-controlled work environments (Szivas *et al.* 2003), and where their social, cultural, educational and skill backgrounds place them in an advantaged position (Iredale 2001; on the links between tourism, migration and entrepreneurship, see Speneger *et al.* 1995). Possessing specialized skills and fulfilling lifestyle-based niche market needs, such individuals find work in the tourism and entertainment sectors of the developing countries alongside native, migrant and immigrant employees who are far less skilled and far less well paid. This sector's 'diverse human capital requirements' (Szivas *et al.* 2003) allow this duality of employment.

The new tourism and entertainment labourscape: an unstructured scene of ethnicization

In the case of Istanbul, this heterogeneity and duality has produced some interesting constellations in terms of tourism and entertainment employment. In certain respects these mirror the patterns observed elsewhere, but they also differ from them. For my research on tourism and entertainment employment in Istanbul, I determined some preliminary points to investigate in these emerging constellations.

I identified three pilot districts to study: Beyoğlu, Sultanahmet and Etiler/ Bosphorus. Beyoğlu (formerly known as Pera), located in the historically most 'Westernized' part of Istanbul, forms the entertainment heartland of the city. With its bars, clubs, restaurants, cinemas, galleries and cultural centres, it caters chiefly to middle-class Turks. Its picturesque historic buildings also make it an attractive destination for tourists, although their numbers are still limited. Sultanahmet, located on the historic peninsula that formerly housed the Byzantine and Ottoman administrative centre, and hence the site of numerous palaces, mosques, churches and other sites, is seen as the tourism stronghold of Istanbul. Etiler/Bosphorus is an important centre for entertainment catering mainly to upper-class Turks; it has little or no tourist presence.[10]

On the basis of my research in these three districts, I would argue that Istanbul's new tourism and entertainment scene has generated a new labourscape in the leisurely service sector.[11] In so doing I am borrowing and extending Appadurai's (1996) notion of various '-scapes' in order to analyse the shifting and fluid labourscape of tourism and entertainment in Istanbul. As this new labourscape becomes constructed and configured, a significant and functional stratifying role is played by a process of bifurcating ethnicization. Istanbul's urban tourist–entertainment– leisure industry employs or incorporates increasingly diverse groups, including many long-time or recent internal migrants, foreign immigrants, Western expatriates, local ethnic minorities, as well as ethnic Turkish local natives. But specific demographic conditions and the unstructured nature of Istanbul's tourism and entertainment setting rule out any one-to-one correspondence with the general pattern discussed in the literature.

The concept of *opportunity structure* (Kloosterman and Rath 2001; Rath 2002a, 2002b) may be employed here to help clarify the situation. The general shift in Turkey towards market-oriented growth; the accompanying relaxation of state regulations on property ownership, business licensing and employment; the expanding share of consumer markets fuelled by well-to-do, globalizing urbanites; and the broad cultural shift towards globalization, diversification and hybridization – all these add up to a time- and space-specific opportunity structure for entrepreneurs and workers in Istanbul. As different groups come to perceive this opportunity structure differently, they mobilize their financial, human and cultural capital in a variety of ways, thereby pursuing different strategies of ethnicization.

In Istanbul, distinct groups of rural migrants, immigrants and local 'ethnics' are employed in specific zones of tourism and entertainment settings. This is because they adapt in different ways to the 'diverse human capital requirements' (Szivas *et al.* 2003) – depending on their social, educational and regional backgrounds and their self-perception of their 'worth' in the setting. Or to put it differently, specific groups, on the basis of their backgrounds, steer towards those parts of the setting where they feel they 'belong', and not necessarily towards areas they are best qualified for. I shall now take a closer look at the position occupied by some of these specific groups in the stratification process.

Accents à la mode: 'chic ethnics'

In line with Istanbul's transformation into a 'globalized' city, many world or ethnic cuisines, as well as world fashion, music and dance styles, have become very much the 'in thing'. As noted above, this cultural shift to globalization is one of the major parameters of the opportunity structure for the employees and entrepreneurs of the tourism and entertainment labourscape in Istanbul. In terms of ethnicized stratification, I would argue that certain foreign staff employed at the top of this bifurcated labourscape form a stratum of 'chic ethnics', occupying conspicuous places at the 'front' of the top tier of the tourism and entertainment labourscape in Istanbul. I am borrowing here Zukin's (1995) typology of 'front' and 'back'. Chic ethnics are highly visible in many establishments, and they are consciously deployed there by the entrepreneurs and owners as a showpiece for their enterprises.

Catering mainly to upper- and upper-middle-class Istanbul residents, as well as to foreign nationals working in Istanbul's globalized sectors, the upscale entertainment venues thus employ high-paid, high-skilled foreigners. Most are citizens of economically advanced Western countries. Western ethnics seem to be the group of preference for the role of chic ethnics in Istanbul's entertainment industry. This relates in particular to the Turkish middle and upper classes' deeply rooted disdain for 'Middle Easterners' (most clearly manifested in occasional anti-Arab sentiments and utterances, which are symptomatic of a strange sort of Turkish

Orientalism or even cultural schizophrenia). It is also explained by the 'Westernized' or 'modernized' consumerist ethic of these classes and by their lack of knowledge about Asian countries.

Istanbul's five-star hotels employ high-profile international chefs, business executives and 'image consultants'. Larger tourism companies take on foreign staff that coordinate activities with Western-based offices and other companies. Some upscale entertainment venues employ occasional or part-time foreign disc jockeys, dance instructors and bartenders as well as foreign design and image consultants. In some cases, employees are attractive to the customers primarily just because they are foreigners. Indeed, the owner of an entertainment establishment I interviewed reported that he had drawn in more customers after hiring an English barman: 'Customers enjoy being served by someone speaking English with an English accent. Well, who would have a more genuine English accent than an Englishman?'

Istanbul's 'night-out', high-society and lifestyle magazines often feature stories about African American blues and jazz musicians, American, Asian and European DJs, Latin dance instructors, British bartenders, international chefs, diet advisers and personal trainers working in that sector. Situated at the top of the sector, they form the crème de la crème of Istanbul's tourism and entertainment world. Istanbul's upper-middle and upper classes thus attach increasing value to the presence of chic ethnics in their consumption patterns, especially when consuming the city's urban, privatized spaces of entertainment.

This attitude is also consistent with Istanbul consumers' increasing acceptance of (or even preference for) foreign language brand names, which creates interesting constellations of heterogeneity, diversity and hybridity. In addition to the growing popularity of American and European fast food, restaurant and café chains such as McDonald's, Arby's, Starbucks, Waffle House and Schlotzky's Deli, some linguistically hybrid food establishments, such as Emmim *(my uncle)* Chicken & Kitchen and Dilim *(slice)* Sandwiches and Cafe have proliferated in recent years. One of the most expensive and upscale Istanbul clubs is called Laila (the English transliteration of an Arab female first name; the Turkish version, Leyla, is the name normally used in Turkey).

Perhaps two creative and humoristic cases deserve mention. A Bağdat Avenue clothing store bears the name Nursace, a takeoff on the famous Versace formed by substituting 'Nur-' *(light)* for the first syllable. And a textile entrepreneur from Rize, a Black Sea coastal town, launched a new line of clothing called Rizelli with the express intention of making his brand 'Italian-sounding'.[12] The Turkish *Rizeli* means 'from Rize', but the added letter *l* makes the brand seem almost Italian!

Tourists-turned-entrepreneurs

Parallel to the advent of chic ethnics as employees, one can also observe the emergence of foreign entrepreneurs peopling the entertainment sector (and to some extent also the tourism sector). The high value now ascribed in Turkey to 'anything foreign' seems to have led Britons, Russians, Spaniards, Latin Americans and Asians to settle in Istanbul and to open restaurants, cafés and other entertainment venues. Many of them once arrived in Turkey as tourists, and decided to stay or return after spotting the potential of the market for their prospective business ventures. Fewer than half of the foreign entrepreneurs I interviewed had any prior training or professional experience in the food or entertainment industries. The national origins of these entrepreneurs (who also qualify as chic ethnics) seemed to overshadow relevant attributes such as educational background and professional skills. This seems in line with Hall and Williams's (2002) finding that many immigrant entrepreneurs set up medium-scale businesses catering to niche markets and/or shaped by lifestyle considerations.

Alongside the sociocultural parameters that make the opportunity struc-ture for such entrepreneurs increasingly favourable (such as the growing predilection for lifestyle-based consumption patterns in Istanbul's middle and upper classes), recent changes in the Turkish legal climate have opened additional incentives to chic-ethnic entrepreneurs and employees. Not only have regulatory structures, such as legal barriers to property and employ-ment rights for non-citizens, been relaxed (and are now being further eased in connection with Turkey's European Union membership bid), but official attitudes towards immigrant entrepreneurs and workers have also warmed as a result of the marketization and privatization trend.

Certain more micro-level decisions by other actors may also awaken perceptions of new business opportunities. After the United States consulate moved from Beyoğlu to Kaplıcalar Mevkii, near İstinye, 'for security reasons', a new leisure-oriented, functional periphery directed at its employ-ees and visitors seems to have emerged in the new neighbourhood, which is a low-income, migrant-dominated and informally settled area. This is illustrated by the emergence of small businesses such as office supply stores, visa advisers, one-hour photo shops, cafés, a Pakistani restaurant and a Dunkin' Donuts franchise. However, as a result of the consulate's 'self-sustaining isolation' and the unwillingness of its employees to leave the premises, many of the new businesses have already had to close down – like the Pakistani restaurateur, who left the neighbourhood reportedly 'without even paying his rent'.[13]

Suitable repertoires to catch the tide: highly presentable locals and visionaries

Concurrently with the arrival of the chic foreign ethnics, local Turkish individuals who are highly educated, highly skilled or of high social standing are now taking part in Istanbul's tourism and entertainment labourscapes too, both as 'front' employees and as entrepreneurs. Turkish nationals with some experience living abroad are a significant group among them.

Many tourism and entertainment establishments hire local knowledge workers with the right 'vision' and 'background'. Many other Turkish citizens of similar backgrounds are launching businesses in tourism and entertainment sectors themselves.

Since few native Turks have educational backgrounds immediately relevant for the tourism or entertainment industries, many new Turkish entrepreneurs have formed a vision for their business ventures while living abroad, developing a cultural repertoire of what is trendy abroad and trying to adapt and apply it after coming back to Istanbul. They also capitalize on other aspects of their experience of living abroad. Some of the entrepreneurs interviewed had come up with interesting new 'themes' while living or studying abroad, and subsequently started businesses in Istanbul catering to new lifestyle considerations. Istanbul's tourism and entertainment industries now spawn new businesses such as bagel cafés, 'finger food' or 'fusion cuisine' restaurants, cultural and historical neighbourhood tours led by archaeologists or anthropologists, and environmental tourism companies. It was interesting to note how frequently the notion of *konsept*, a Turkified version of the English word *concept*, was used by entrepreneurs in characterizing their businesses, even though a Turkish equivalent *kavram* was readily available but was not used at all.

Such individuals somewhat fit the category of what some analysts call White Turks *(Beyaz Türkler)*, indicating a leaning towards an entrepreneurial, individualistic, neo-liberal, self-reflexive and consumerist ethic, as well as a more privileged, upper-class and urban social background. The classification is based on a social and cultural construction rather than on biological differences (although lighter skin does seem to be preferred by some people, and has been used as a further tool of stratification). In popular iconography, as well as in the language of printed and television advertising, White Turks are often counterposed to uneducated, crude and darker Turks. Initially used in a pejorative sense by their critics, the notion of White Turks has more recently been incorporated into a more celebratory discourse by some neo-liberal, Western-oriented analysts and newspaper columnists. Notably, some of the interviewed entrepreneurs also unapologetically and self-reflexively used the label to define both themselves and the 'target audiences' or 'niche consumers' for their businesses.

This picture seems to conform to the general pattern of correlation between high qualification and high pay. What distinguishes Istanbul in this respect is that this correlation can be observed between privileged social backgrounds and high pay, as well as between high qualifications and high pay. In other words, individuals from high social backgrounds, who may or may not be sufficiently qualified, occupy many of the high-paid jobs. Many of them deploy their cultural or social capital to become and remain successful within the top echelons of the tourism and entertainment setting in Istanbul.

Ethnicization at the bottom

The other prong of the bifurcation process involves real, imagined or constructed ethnicization in the lower levels of Istanbul's tourism and entertainment labourscape. Internal migrants, documented and undocumented immigrants, native locals and long-time 'minority' residents of Istanbul all find employment in mainly low-paid jobs. For instance, ethnic Turks, ethnic Kurds, recent immigrants from the former Socialist bloc, as well as Roma (who have historically constituted an informal entertainment labour force) now work in establishments as caterers, entertainers and support staff. Although most such jobs require low qualification and human capital, that is not necessarily the case for all the jobs concerned.

To understand this side of the stratification process, one needs to identify three additional, more micro-level substratification strategies. The first amounts to steering individuals with lower skill and educational levels towards low-paying jobs. It is mainly the migrants from Anatolia, including ethnic Kurds, that populate this substratum. Such support staff are usually invisible and readily replaceable, and are concentrated at the back of the bottom tier of the tourism and entertainment labourscape. As dishwashers, cooks and cleaners, they are the silent labour force behind the visible scene. Ethnicity is a functional factor here inasmuch as it helps channel the workers to the bottom and back regions of the labourscape (especially if it involves shakier social roots). These workers' ethnicity is not for display.

Although the workers' replaceability is generally very real, an interesting clustering and concentration seems to occur among cooks in particular. Most of the cooks I interviewed were recent migrants from Mengen, a small town near the Anatolian city of Bolu, well known for the culinary talents of its male population. Migrants from Mengen seem to dominate the kitchen scene almost universally, and when a Mengen cook is fired or resigns, informal networking practices quickly help find another Mengen cook to replace him. Although Mengen's cooks are best known for their skills in 'local', Anatolian cuisine, some international cuisine restaurants and bagel cafés also employ them. In those cases, the Mengen roots of the cooks are, unsurprisingly, not emphasized in promotional literature. Not all Mengen cooks are situated in the bottom tier, as many renowned and highly skilled Mengen 'chefs' (as opposed to 'mere' cooks) earn high incomes at more established, expensive restaurants and hotels.

In a second substratification strategy, pursued especially in tourism, some employees are deliberately ethnicized for display purposes. Conspicuously exhibited 'folkloric' musicians and dancers, as well as kebab restaurant employees wearing 'traditional' outfits, help construct an image of the exotic Other as service provider for consumption by tourists. They are placed at the 'front'. Most of the Sultanahmet-based hotels and tour companies routinely offer 'visits' to 'authentic' restaurants or music and dance shows, where the staffs are required to wear 'representatively authentic' clothing items, including some that are seriously outdated or even legally banned, such as the fez and the turban. An almost generic, retroactive 'ethnicity' is constructed, which erases local distinctions. The idea is to display a blanket 'Turkishness', and to camouflage away the real ethnic differences within Turkey – especially the problematic ones. Any reference to Kurds is carefully avoided, for example, but also any acknowledgement of the non-Turkish roots of the fez.

Turn up your ethnicity

A third strategy, again at the 'front', further highlights the ethnic backgrounds of certain employees. This notably applies to Roma workers and to immigrants from the former Socialist bloc. The Roma, historically serving as an informal labour force in entertainment and tourism and employed as full-time or occasional music performers and dancers, have seen their ethnicity emerge into the foreground in recent years. Still located near the bottom tier of the stratification, but at the front, Roma workers are highly skilled, but earn low pay, mainly because of their persisting informal employment terms and shaky legal status.

Popularly labelled as *çingene* (gypsy) or, even more pejoratively, *çingen*, Roma people have not only been openly disdained and ridiculed, and their Islamic self-identification called into doubt, but since the Ottoman period they have also had to endure a precarious status as *esmer vatandaş* (swarthy citizens). In 1934, in line with global legal measures designed to limit Roma mobility, çingene were banned from entering Turkish territory. Even a recent, avowedly liberal naturalization law, designed to facilitate EU accession, lists 'unethical behaviour' and 'working as a çingene' as grounds 'making applicants ineligible' for Turkish citizenship.

The current heightened visibility of the otherwise disdained Roma ethnicity results not only from a long-term struggle by Roma rights activists, but also from the recent trendiness of 'fusion' as a musical and cultural form. Roma musicians – previously not taken seriously in the music and entertainment business beyond their role as informal performers mainly dependent on customers' tips – have now been discovered, bringing them a new respectability. Albums or concerts are now common which bring together groups like the Amsterdam Klezmer Band and the Galata Gypsy Band, or the Brooklyn Funk Essentials and the Roma group Laço Tayfa. The concerts are often staged in Istanbul's cutting-edge, experimental new clubs like Babylon. Some larger hotels in Sultanahmet have helped to release albums by the Ahırkapı Roman Orkestrası and to promote the orchestra as a signature item in hotel entertainment shows. A new 'gaze' towards Istanbul's 'own ethnicities', in a capacity as musical subjects, may be said to have developed in the city's new spaces of music and entertainment. Within this atmosphere, Roma 'ethnicity' becomes highly commodified, while social and legal exclusionary practices and the residential and occupational segregation of Roma people still continue.

Certainly an important question remains as to whether these experimental, fusion-oriented musical efforts, and the spaces of music they create, substantively bridge the social distance between the entertainment economies of Beyoğlu and Maslak (where many crossover-promoting establishments are housed) and other areas like Tarlabaşı or Ahırkapı (lower-income neighbourhoods out of which many Roma musicians are recruited to play with Western bands). Tarlabaşı, for example, is just a stone's throw from Beyoğlu, and its mainly Roma and Kurdish residents are vitally, yet very informally, connected to Beyoğlu's entertainment economy as street vendors, street musicians and fortune tellers. At the same time, media accounts and popular attitudes also continue to blame both groups for Beyoğlu's criminal economy, citing activities like drug dealing, 'mafia' car parking practices and street robbery networks. Living under the shadow of criminalization, many Tarlabaşı residents may not be enjoying the fruits of the new musicscapes in the same way as their friends or relatives who were recruited to play at Babylon (Aytar and Keskin 2003).

Eroticized Others

A final group whose members have been effectively ethnicized consists of mostly undocumented immigrants from the former Socialist bloc. Having flooded to Turkey prior to and since the breakup of the Soviet Union and other Socialist regimes, many of these immigrants pursue informal 'suitcase' trading – buying up Turkish consumer goods from local merchants and selling them back home at a profit, or setting up street bazaars in Turkish cities to sell Socialist-era paraphernalia and other cheap, 'interesting' or hard-to-find items. Peopling the lively transnational informal trading networks,[14] they have helped Istanbul's Laleli neighbourhood to become a major node of that trade.

Some of these undocumented immigrants have also become caught in a web of sex trade. Hundreds of undocumented Eastern and Central European and Central Asian women now reside in Istanbul as informal sex workers or female escorts at entertainment venues such as the *pavyon* or the *taverna*. Many of them have college or even postgraduate degrees and would be well qualified for other, more formal employment. They may indeed occupy somewhat privileged positions on the catering and entertainment ladder, but cannot get highly paid employment because of their legal status. Because their 'exotic' and eroticized Otherness is their vital resource for employment, they are effectively 'ethnicized' (as well as being criminalized and subjected to patriarchal and semi-official violence). Labelled with pejoratively used blanket terms such as *Rus* (Russian) or *Nataşa* (Natasha), they constitute yet another segment of the entertainment and tourism labourscape in Istanbul – a segment that is at the front, but not necessarily in the top tier of the spectrum.

Conclusion

With the turn towards a market-oriented growth strategy and receding state interventionism and regulation in Turkey since the 1980s, Istanbul has become a 'globalizing' city. A new stratum of well-to-do urbanites – employing consumption as a self-reflexive, social boundary-setting activity – has been the major engine behind the cultural shift to globalization, diversification and hybridization. These factors have jointly paved the way towards a conducive opportunity structure for entrepreneurs and workers in a new entertainment and tourism setting.

The new demographic, social, economic and cultural morphology of Istanbul has generated demand for a wide spectrum of goods and services. The emergent, yet still rather unstructured and ambiguous, demand created by this opportunity structure was ripe to be further moulded – and variously customized – by a growing group of entrepreneurs and staff, who have adapted themselves to the new environment and have deployed their social, human and cultural capital in diverse ways.

A strategy of social bifurcation based on ethnicization, which includes several micro-level substratification strategies, is revealed in the ways in which the entrepreneurs and employees have differently positioned themselves in this new, fluid environment of demand and opportunity. Ethnicization has gradually created a pattern of stratification for those incorporated into or employed in Istanbul's new tourism and entertainment setting. The ethnic backgrounds of selected groups of people have been highlighted and functionalized in this ethnicization process, while the backgrounds of some others have been camouflaged. Ethnicization in Istanbul has not yet proceeded so far as to generate ethnic tourist enclaves (residential segregation notwithstanding), as has occurred in some more Western cities. Yet it has already become a significant force in the commodification of tourist and entertainment spaces and experiences, as well as in the peopling and staffing of a 'lifestylized' Istanbul.

However different this process may be from the postindustrial experiences of more advanced economies, Istanbul's new entertainment and tourism environment nevertheless presents an informative case study. It shows how a new service sector can spring up within an opportunity structure shaped mainly by new macroeconomic and political realities and by a permissive sociocultural atmosphere of consumerism, all of which facilitate market diversification. The development of this sector also highlights the importance of the agency of the entrepreneurs and workers in simultaneously shaping this new environment – thereby also shaping themselves and their various forms of capital. It is not yet clear how long this opportunity structure, and the agency of the entrepreneurs and workers within it, will last. That will be tested in the future of this commodified and catered metropolis of consumption.

Notes

1 The research on which this paper is based was partially funded by the
  MEAwards Program in Population and the Social Sciences and by the
  Population Council's West Asia and North Africa Regional Office in Cairo.
2 Quoted from a personal communication with an Istanbul resident. The
  Turkish singer Sertab Erener won the 2003 Eurovision Song Contest with
  her English-language song, *Everyway that I can*. The accompanying music
  video and her stage performance at the contest were pervaded by Orientalist
  images of belly dancers. Later she provoked controversy by releasing a new
  music video of a song with English lyrics, *Here I am*, showing homoerotic
  images of Turkish oil wrestlers.
3 During the 1980s, a 20 per cent shift occurred in the Turkish rural:urban
  population ratio (Ayata 1997).
4 In the mid-1970s, almost half of all manufacturing enterprises in Turkey were
  based in Istanbul, claiming more than half of the manufacturing employment
  in the city (Özmucur 1976).
5 Ninety-five per cent of the producer service firms receiving foreign capital
  have been established since 1984, and nearly 75 per cent of these have located
  in Istanbul (Erkip 2000; Tokatlı and Erkip 1998). In this new environment,
  foreign investors prefer to invest in non manufacturing areas such as tourism
  and producer services (Erkip 2000).
6 Keyder and Öncü (1994) have argued that Istanbul is a 'globalizing city'
  rather than a 'global' (or 'world') city as usually conceptualized in the literature
  (see also Keyder 1993, 1999; Öncü 1997).
7 Parallel to the turn towards a market-oriented strategy, the former depend-
  ence on simple mass tourism was shelved, and many forms of tourism
  in Turkey, including urban tourism, became more diversified (for various
  discussions, see Göymen 2000 and Tavmergen and Oral 1999; on the proliferation, spatial patterning and diversification of Istanbul hotels, see
  Dökmeci and Balta 1999). The latter article also usefully identifies connections between older forms of cultural and historical tourism, the newly
  flourishing business and convention tourism, and the proliferation and
  diversification of hotels.
8 Pérouse (2000: 160) has argued that as a result of the recently intensified
  orientation of Turkey towards Europe, Istanbul increasingly occupies a 'key
  function in the active interface of the international system of migration'. Not
  only does it play host to transit immigrants and refugees (Iraqis, Pakistanis,
  Sri Lankans, Nigerians, Kurds and others) en route to Europe, but it also houses
  more permanent or semi-permanent foreign residents. Among the latter groups,
  Pérouse lists Iranians, immigrants from the former Socialist bloc, and Western
  European professionals, particularly Germans (totalling around 6500 in
  Turkey in 1998). Increasing immigration from Africa is a significant, but seriously under-researched phenomenon. Especially in the light of developments
  such as these, Pérouse and others have called for more research on this 'new
  cosmopolitanism' in Istanbul.
9 See Zukin (1995) for a view of a similar situation in New York's restaurants,
  where a comparable duality exists in the service sector. For discussions of
  other types of service sector employment, see Bailey (1985); Sassen (1996,
  1998); Waldinger (1992).
10 By 'Bosphorus' I am mainly referring to the slice of the Bosphorus shores
  between Istanbul's two suspension bridges. Etiler is a high-income neighbourhood on the European side of the Bosphorus. The 'Etiler–Bosphorus line'
  is referred to in daily life and in the media as Istanbul's 'pleasure and

entertainment jewel'. Kuruçeşme, Bebek and Tarabya are Bosphorus neighbourhoods with upscale entertainment establishments similar to those in Etiler.

11 My research was conducted from July to October 2003. A total of 12 owners, 24 employees and 24 customers of 12 establishments underwent semistructured interviews. I also carried out a media review from July 2003 to June 2004, scanning accounts in monthly, weekly and daily publications: *Time Out Istanbul, Istanbul Life, Zip Istanbul* (monthly), *Beyoğlu* (a 'what-to-do' weekly) and *Radikal, Milliyet, Hürriyet* and *Sabah* (major daily newspapers carrying frequent reviews of establishments). I further consulted Istanbul tourism and hotel guides and other tourism publications.

12 B. Eren, 'Yabancı markalarımızı artik kendimize "uyduruyoruz"' [We now 'adapt' foreign brands to ourselves], *Zaman*, 19 October 2003.

13 T. Soykan, 'Amerikan rüyasi çabuk bitti' [The American dream fades real quick], *Radikal*, 27 October 2003.

14 For a good discussion of this informal transnationalism from below, see Yükseker (2000).

References

Aitken, C. and Hall, C.M. (2000) 'Migrant and foreign skills and their relevance to the tourism industry', *Tourism Geographies*, 2(1): 66–86.

Akşit, B. (1998) 'İçgöçlerin nesnel ve öznel tarihi üzerine gözlemler: köy tarafından bir bakış' [Observations on the objective and subjective history of the internal migrations: a look from the village side], in *Türkiye'de İçgöç* [Internal Migration in Turkey], Istanbul: Tarih Vakfi.

Akşit, B., Mutlu, K., Nalbantoglu, H., Aksay, A. and Sen, M. (1996) 'Population movements in southeastern Anatolia: some findings of an empirical research in 1993', *New Perspectives on Turkey*, 14: 53–74.

Aksoy, A. (1996) *Küreselleşme ve İstanbul'da İstihdam* [Globalization and Employment in Istanbul], Istanbul: Friedrich Ebert Foundation.

Appadurai, A. (1996) *Modernity at Large: cultural dimensions of globalization*, Minneapolis: University of Minnesota Press.

Arıcanlı, T. and Rodrik, D. (eds) (1990a) *The Political Economy of Turkey: debt, adjustment and sustainability*, Basingstoke: Macmillan.

Arıcanlı, T. and Rodrik, D. (1990b) 'An overview of Turkey's experience with economic liberalization and structural adjustment', *World Development*, 18: 1343–50.

Ayata, A. (1997) 'The emergence of identity politics in Turkey', *New Perspectives on Turkey*, 17: 59–73.

Aytar, V. and Keskin, A. (2003) 'Constructions of spaces of music in Istanbul: scuffling and intermingling sounds in a fragmented metropolis', *Géocarrefour*, 78(2): 147–58.

Bailey, T. (1985) 'A case study of immigrants in the restaurant industry', *Industrial Relations*, 24(2): 205–21.

Bourdieu, P. [1979] (1984) *Distinction: a social critique of the judgement of taste*, trans. R. Nice, Cambridge MA: Harvard University Press; originally published as *La Distinction: critique sociale du jugement*, Paris: Editions de Minuit.

Carr, N. (2002) 'The tourism-leisure behavioural continuum', *Annals of Tourism Research*, 29(4): 972–86.

Çelik, A.B. (2002) 'Migrating onto identity: Kurdish mobilization through associations in Istanbul', MA thesis, State University of New York at Binghamton.

Danielson, M. and Keleş, R. (1984) *The Politics of Rapid Urbanization: the government and growth in modern Turkey*, New York: Holmes and Meier.

Dökmeci, V. and Balta, N. (1999) 'The evolution and distribution of hotels in Istanbul', *European Planning Studies*, 7(1): 99–109.

Erkip, F. (2000) 'Global transformations versus local dynamics in Istanbul: planning in a fragmented metropolis', *Cities*, 17(5): 371–7.

Göymen, K. (2000) 'Tourism and governance in Turkey', *Annals of Tourism Research*, 27(4):1025–1048.

Hall, C.M. and Williams, A.M. (2002) *Tourism and Migration: new relationships between production and consumption*, Dordrecht: Kluwer Academic.

Iredale, R. (2001) 'The migration of professionals: theories and typologies', *International Migration*, 39(5): 7–24.

Kandiyoti, D. (2002) 'Introduction: reading the fragments', in D. Kandiyoti and A. Saktanber (eds) *Fragments of Culture: the everyday of modern Turkey*, New Brunswick NJ: Rutgers University Press.

Karpat, K. (1976) *The Gecekondu: rural migration and urbanization*, Cambridge: Cambridge University Press.

Keyder, Ç. (1987) *State and Class in Turkey*, London: Verso.

Keyder, Ç. (1993) *Ulusal Kalkınmacılığın İflası* [The Bankruptcy of National Developmentalism], Istanbul: Metis.

Keyder, Ç. (ed.) (1999) *Istanbul: between the global and the local*, Lanham MD: Rowman and Littlefield.

Keyder, Ç. and Öncü, A. (1994) 'Globalization of a third-world metropolis: Istanbul in the 1980's', *Review*, 17(3): 383–421.

Kiray, M.B. (1998) *Kentleşme Yazilari* [Writings on Urbanization], Istanbul: Baglam.

Kirişçi, K. (1998) 'Turkey', in J. Hampton (ed.) *Internally Displaced People: a global survey*, London: Earthscan.

Kloosterman, R. and Rath, J. (2001) 'Immigrant entrepreneurs in advanced economies: mixed embeddedness further explored', *Journal of Ethnic and Migration Studies*, 27(2): 189–202.

Li, F.L.N, Finlay, A.M. and Jones, H. (1998) 'A cultural economy perspective on service sector migration in the global city: the case of Hong Kong', *International Migration*, 36(2): 131–50.

Navaro-Yashin, Y. (2002) 'The market for identities: secularism, Islamism and commodities', in D. Kandiyoti and A. Saktanber (eds) *Fragments of Culture: the everyday of modern Turkey*, New Brunswick NJ: Rutgers University Press.

Öncü, A. (1997) 'The myth of the "ideal home" travels across cultural borders to Istanbul', in A. Öncü and P. Weyland (eds) *Space, Culture and Power: new identities in globalizing cities*, London: Zed Books.

Önder, N. (1997) 'International finance and the crisis of neoliberal economic strategy in contemporary Turkey', *Journal of Emerging Markets*, 2(3): 21–56.

Özmucur, S. (1976) 'İstanbul ili gelir tahminleri' [Income estimates for the province of Istanbul], *Bogaziçi Üniversitesi Ekonomi Dergisi*, 4/5: 51–65.

Özmucur, S. (1995) *Yeni Milli Gelir Serisi ve Gelirin Fonksiyonel Dağılımı Geçici Tahminleri, 1968–1994* [New Series on National Income and Temporary Estimates on the Functional Distribution of Income, 1968–1994], Istanbul: Boğaziçi Üniversitesi Araştirmalari.

Pérouse, J.F. (1999) 'Istanbul, capitale du nouveau monde turc?' *Revue Française de Géoéconomie*, 9: 45–53.

Pérouse, J.F. (2000) 'L'internationalisation de la métropole stambouliote', in *Méditerranée et Mer Noire entre Mondialisation et Régionalisation*, Paris: L'Harmattan.

Rath, J. (2002a) 'Do immigrant entrepreneurs play the game of ethnic musical chairs? A critique of Waldinger's model of immigrant incorporation', in A. Messina (ed.) *Continuing Quandary for States and Societies: West European immigration and immigrant policy in the new century*, Westport CT: Praeger.

Rath, J. (ed.) (2002b) *Unravelling the Rag Trade: immigrant entrepreneurship in seven world cities*. Oxford and New York: Berg.

Sassen, S. (1996) 'New employment regimes in cities: the impact on immigrant workers', *New Community*, 22(4): 579–94.

Sassen, S. (1998) *Globalization and Its Discontents*, New York: The New Press.

Şenses, F. (1990) 'Türkiye'de gelir dağilimi, gelirin yeniden dağitimi ve işgücü piyasalari' [Income distribution, income redistribution and labour markets in Turkey], Working Paper ERC/19903, Ankara: Economic Research Center, Middle East Technical University.

Sönmez, M. (1996) *İstanbul'un İki Yüzü: 1980'den 2000'e değişim* [Two Faces of Istanbul: change from 1980 to 2000], Ankara: Arkadaş Yayinlari.

Speneger, D.J., Johnson, J.D. and Rasker, R. (1995) 'Travel-stimulated entrepreneurial migration', *Journal of Travel Research*, 34(1): 40–51.

Szivas, E., Riley, M. and Airey, D. (2003) 'Labor mobility into tourism: attraction and satisfaction', *Annals of Tourism Research*, 30(1): 64–76.

Tavmergen, I.P. and Oral, S. (1999) 'Tourism development in Turkey', *Annals of Tourism Research*, 26(2): 449–51.

Tokatli, N. and Erkip, F. (1998) 'Foreign investment in producer services: the Turkish experience in the post-1980 period', *Third World Planning Review*, 20(1): 87–106.

Waldinger, R. (1992) 'Taking care of guests: the impact of immigrants on services: an industry case study', *International Journal of Urban and Regional Research* 16: 97–113.

Yenal, N.Z. (2000) 'The culture and political economy of food consumption practices in Turkey', PhD thesis, State University of New York at Binghamton.

Yenal, N.Z. (2003) 'Rüzgàr bizi nereye götürecek?' [Where will the wind take us?], *İstanbul*, 45: 106–9.

Yükseker, H.D. (2000) 'Weaving a market: the informal economy and gender in a transnational trade network between Turkey and the former Soviet Union', PhD thesis, State University of New York at Binghamton.

Zukin, S. (1995) *The Cultures of Cities*, Cambridge MA and Oxford: Blackwell.

6 Immigrants, tourists and the metropolitan landscape

Producing the metropolis of consumption in Orlando, Florida

Hugh Bartling

In March 2002, Mayor Glenda Hood of Orlando, Florida, gave her annual state-of-the-city speech to an assembly of local politicos, real-estate developers and local celebrities. Before extolling the virtues of her municipal stewardship over the course of three-and-a-half terms in office, she revealed her vision of Orlando. Hood first observed that:

> there are few names that cross international barriers, ... names that know no race, color or economic status. Coca Cola is one such name. McDonalds is another. From Fargo, North Dakota, to Osaka, Japan, and everywhere in-between, these names resound in an international language.

She then made the connection:

> Orlando is another of those few names that bring recognition and smiles to faces all over the world. The global brand of Orlando was created by the success of our tourism industry. They set the engine in motion to transform a mid-sized traditional city into the number one tourist destination in the world and established Orlando as a prime location to do business.
>
> (Hood 2002)

Conceiving the city as a 'brand', Hood went on to argue that the image of Orlando was now in need of updating. As a civic booster, she was reluctant to speak openly of the dire economic straits in which the tourist industry found itself in the aftermath of the 11 September 2001 terrorist attacks. A sharp drop in the numbers of national and international visitors had triggered massive job cuts. Hood's suggestion was to reinvigorate the old city slogan 'the City Beautiful' by superimposing the descriptor 'new' – thus rebranding Orlando as the 'New City Beautiful'.

While the merits of the new branding and the creative acumen of Mayor Hood may remain the subjects of debate, her complete and explicit embrace of a 'branding ethic' speaks to a larger trend in urban development, both in the Central Florida region and throughout the United States.[1] Promoting the 'globally competitive' city is the new mantra of local elites and boosters as they search for ways to distinguish regions and promote local economic development in an environment marked by deindustrialization and the uncertainty of prolonged private investment. The structural economic transformations that animate such promotions are very real phenomena. Federal disinvestment in urban areas, coupled with corporate capital's increasing ability to transcend space, has made a reliance on affective strategies almost inevitable. A carefully constructed 'image' or 'brand' thereby serves as an emotional signifier, intended to engender attachments to a particular place. In this post-industrial era and informational economy, the city's branded aura is its greatest asset. In the rush to 'brand' the city, an enormity of effort goes into creating not only a saleable image, but also a corresponding landscape to validate the brand.

The purposeful definition or 'spin' of the city is not necessarily a new phenomenon. What is significant in our contemporary era are the particular social and political practices that accompany these efforts at branding, as well as the purposeful disconnection made between the practices of cultural representation and the lived experiences of many individuals who are actually involved in the material production of these commodified spaces.

By virtue of its position as one of the world's most popular vacation destinations, and because of the central role that 'brand building' plays in the effort to distinguish this particular metropolis in the world's vacationing psyche, Orlando is an important site in which to examine the role that image construction and representation play in informing the lived experience of the tourist metropolis. Far from describing the consumptive experiences of the carefree tourist, I will focus here on the dynamic relation between consumption and labour, with particular attention to the role of immigrants and migrants in this process. The globalized machinery of the region's consumptive dynamic is often obscured by Orlando's worldwide affective appeal and by the eagerness of elite forces to exploit the tourist sentiment. The processes whereby immigrant and migrant labour influence the nature of the tourism landscape in the tourist economy are varied and uneven. This is because of the multifunctional characteristics of service sector employment and the wide variance in worker interaction with the public.

In this chapter I analyse Orlando as a metropolitan region seeking to assert its attractiveness as a commodity form. Building from the theories of Marx and Lukács on the mystical character of commodities, I map the various ways in which the immigrant presence informs the production of the spaces of consumption in Orlando's tourist region. In particular I highlight the bifurcated nature of immigrant labour in the Orlando region: immigrants and migrants from Western countries are accorded a public presence in hospitality work, while their counterparts from the

Global South largely serve functions behind the scenes. Despite the uneven nature of the immigrant presence, however, the various immigrant and migrant communities have certain commonalities in the metropolitan tourist landscape. I will point out some ways in which a new immigrant politics is emerging, linking local concerns with national and international struggles.

Consumption and the city-as-commodity

The ascendancy of post-industrial economies has fostered an ethic of consumption and commodification. Thirty years ago, Harry Braverman (1974: 281) analysed the tendency of the capitalist mode of production to expand to the point where 'the inhabitant of capitalist society is enmeshed in a web made up of commodity goods and commodity services from which there is little possibility of escape except through partial or total abstention from social life as it now exists'. Braverman was focusing in particular on the expanding number and scope of consumer goods that have accompanied industrialization and post-industrialization. As expansive capitalism brings more and more aspects of human life under its purview, needs are developed that can only be met by appropriate commodities. These, in turn, generate new needs and new commodities, thus fuelling the never-ending cycle of commodity expansion.

While the practice of commodifying urban spaces has a long history, the notion of the city as a product has taken on special importance in recent years. As cities retool themselves in the face of deindustrialization, a need is felt to construct the understanding of the city as a particular type of commodity. The contemporary trend in urban branding can be read as a further step in the commodification processes. Metropolitan elites, no longer able to compete globally in manufacturing or other sectors where capital can be disinvested with ease, now regard the construction of the 'city-as-product' as a way to mitigate excessive outward capital flow. For if the city itself is the commodity, the economic relations surrounding its production cannot be exported elsewhere.

The processes and policies that support this branded image of the city can easily come into conflict with other important social and economic needs of complex metropolitan regions. Because a commodity's value in the age of the informational economy increasingly derives from the image or aura that is created around it, this metropolitan 'image production' usually has to hide away a variety of ugly material inequities which are essential to sustaining the commodity's existence.

As Marx and his intellectual heir, Georg Lukács, both made clear, the process of producing a commodity is rife with contradictions and mystifications (see e.g. Lukács [1923] 1972, Marx [1890] 1992). In this sense, the branded commodity is a selective representation of the social world, often with elimination of any reference to essential processes in its production. The commodity obscures the social relations of production, thus occluding the view of fundamental structural conditions that could provide a fuller understanding of the machinations of a globalizing tourist industry in a localized place.

Since this commodity form embodies and encompasses the urban landscape, the imprints of this branding of the metropolitan region are significant because they reflect powerful frameworks that guide social and political relations. This particularly applies to workers in the tourist industry. Tourism is a labour-intensive industry whose functioning depends on the smooth completion of a large number of disparate tasks. As with all commodities, the allure of the object of production is rarely represented as a product of social labour. The product's attractiveness derives from an appeal to affective longings such as escape, relaxation, adventure and excitement. A key difference between tourist commodities and other objects of production in terms of labour lies in the spatial commingling of production and consumption. The actual product of tourism is essentially spatial, meaning that the consumptive experience is bounded. The touristic space is specifically designed to induce a certain type of consumptive behaviour. The effective promotion of consumptive spaces, in turn, requires a consistent productive regimen which is spatially coterminous with that of consumption. Hence, tourist spaces are simultaneously areas of production and consumption, operating as functional spatial commodities, yet requiring a distinct dialectical separation between representation and production.

Representation, here, is the process of creating a desirable space that attracts tourists. Tourist space, by its very nature, must have a representative aura that simultaneously invokes a sense of uniqueness, departing from the banalities of everyday life, but also a sense of familiarity and/or safety. The novelty must not be excessive, for the space might then spill into the realm of uncertainty or even hostility – a factor that could threaten the profit realization processes that animate the structural realities of tourism in a capitalized environment.

The dialectic of tourist space production especially reveals itself in the maintenance of particular representations. The process of defining and representing tourist spaces entails negotiating the often conflictive requirements for simultaneous consumption and production. As we shall see, this dialectical process is reflected in numerous aspects of the immigrant presence.

Orlando and the metropolis of consumption

The cultural landscape of Orlando is an especially fruitful environment to witness the divergent trends in commodity production. Orlando as a 'global metropolis' owes its contemporary existence to trends initiated during the post–World War II era of mass consumption. Major growth in the region accompanied its incursion into the informational commodity market, which began with the establishment of the Walt Disney Company's first East-coast outpost in 1967.

For an entrepreneur like Disney, the commodity form of 'amusement' (broadly defined) proved pliable enough to attach to various spaces and objects of consumption, as evidenced by his first development in Anaheim, California. In retracing Disney's initial incursion into Central Florida, Richard Foglesong (2001) has exposed the complicity of local elites in enabling the company to acquire land cheaply. Disney, flush from the successes of Disneyland, sought expansion in the mid-1960s, and the company launched a rather clandestine effort to identify both a potential market and a place where land was relatively inexpensive. The *Orlando Sentinel*, the daily regional newspaper, noticed in land purchasing records that large parcels of land were being bought up. According to Foglesong, the publisher also had evidence that Disney was behind the acquisition, and even went so far as to assist the company's buyers in persuading a recalcitrant landowner to sell. Throughout this process in 1965, however, the newspaper and other local elites kept silent on the Disney presence, until the company had acquired enough land to build its second theme park.

The opening of Orlando's Walt Disney World in 1971 indeed had a tremendous impact on the region, spurring rapid population growth and identifying Orlando both with Disney and as a tourist destination with national, later global, appeal. Although the development ignited an immediate round of competing tourist development – most notably with the opening of Sea World in 1973 – the growth of the region did not reach exponential proportions until the 1980s. Walt Disney World and Sea World remained the two dominant attractions in the region until Disney added its Epcot centre in 1982.

The emergence of Epcot prefaced a nearly two-decade period of vigorous growth in the Central Florida tourist industry. During the 1980s, this expansion framed the rise of the consumptive metropolis. The whole texture and existence of this metropolis has been predicated on a major immigrant and 'global' contribution, which both furnishes cultural 'material' for the substantive 'themes' of the Orlando tourist experience and supplies the very material labour to ensure the day-to-day functionality of the industry.

Immigrants, migrants: producing the consuming metropolis

Orlando exemplifies an immigrant dyadic whereby the migratory global Other constitutes the 'cultural material', on the one hand, and the materiality of 'cultural production' on the other. In both cases, the immigrant presence is an essential variable in constructing the cosmopolitan tourist metropolis. Yet the ways in which that presence is articulated, represented and ordered are highly disparate, and speak to the variability inherent in the processes of deterritorialization which accompany the wider commod-ification of the urban form. To understand the tourist metropolis as a space for the commodified experience of everyday life, we must recognize the multiple levels at which the realms of production and representation are interrelated.

Orlando and the surrounding Orange County have been destinations for immigrant workers for decades. Prior to the ascendancy of tourism as the region's major economic force, immigrants were attracted by oppor-tunities in the citrus industry. Orange and grapefruit groves were planted by early settlers in the late nineteenth century, and were first oriented towards domestic and regional consumption. Much of the early citrus workforce was African-American, but after postwar advancements in food processing and long-distance distribution networks created a boom in the industry, the demand grew for temporary workers during peak harvest times. The Minute Maid orange juice brand had its origins in Central Florida, and by the 1970s, after the company was sold to the multinational Coca-Cola Company, the region was attracting significant numbers of migrant agricultural workers – predominantly from Mexico and the Caribbean. Workers would traditionally follow the harvest of citrus and other fruits and vegetables from southern Florida through the Orlando region northwards to Georgia and the Carolinas, getting paid according to the amount of produce they picked. As the volume of production grew further during the 1970s and 1980s, many Mexican, Central American and Haitian families established permanent (though often undocumented) residence in the area.

In a temporal sense, the settling of large numbers of permanent undoc-umented immigrants in Orlando coincided with the expansion of the tourist industry. Interviews with representatives of immigrant rights organizations suggest that tourism industry opportunities emerged during the early mid-1980s for many Latino immigrants, just as unionization efforts among the area's citrus pickers were meeting increasing resistance.

One of the oldest immigrant rights groups in the region is the Farmworker Association of Florida (FWAF), which emerged after the United Farm Workers (UFW) successfully negotiated a contract for citrus pickers with Coca-Cola in the early 1970s. FWAF's current general coordinator, Tirso Moreno, was involved with UFW's Coca-Cola organizing efforts; he had hoped that the union would dedicate additional resources to organizing smaller groves. When the UFW failed to expand its Florida presence in the early 1980s, Moreno and his colleagues established the FWAF as an advocacy and nascent production group. The organization – based in the north-western Orange County community of Apopka – developed immediate social service programmes, such as a health clinic, a cooperative grocery and adult education.

Significant regional economic shifts in the early 1990s created different opportunities and challenges for immigrant workers. First, the citrus sector fluctuated heavily as a consequence of damaging weather conditions and increasing global trade. While the period between 1987 and 1997 saw a statewide increase in the fruit-bearing acreage dedicated to orange production from 375,400 to 624,000 acres (150,000–250,000 ha), much of that growth took place further away, in the south-west part of the state (Agricultural Statistics Board 1999). The Upper Interior citrus-producing district of Florida, which includes Orange County, saw its share of state citrus production fall from 26 per cent in 1982 to 6 per cent in 1999 (Hodges *et al.* 2001).

One of the pressures facing the citrus industry was the vast exurban sprawl, which made property more valuable for development than for continued agricultural growth. As an indication of the shift from agricultural land use toward development, the monetary output of Orange County's environmental horticulture sector – which comprises the development-related activities of landscape maintenance – grew by 48 per cent while fruit and vegetable output remained stagnant (Hodges and Mulkey 2003).

These shifts had impacts on the experiences of many Latino and Caribbean migrant workers. According to Moreno, migrants were finding less employment in agricultural sectors, and more in supportive occupations for the new development and tourist industries.[2] Expansion in the major tourist destinations – Disney, Universal and the Orange County-owned Convention Center – spawned a boom in construction and the attendant service jobs. Unlike metropolitan areas in the Midwest or the north-east United States, which have a long, entrenched skilled trade union presence in the construction industry, Florida has had a largely union-free construction workforce. Because of this, and because much of the commercial and residential construction relies on subcontracting – making it easy for large corporate developers to avoid liability for using undocumented workers – many immigrants from the Latino populations that tended to dominate citrus picking now shifted to construction work.[3]

The importance of the construction industry and the labour involved must not be underestimated if we are to understand the functioning of the tourist metropolis. On the most basic level of erecting tourist spaces of consumption, immigrant labour plays an indispensable role. Low wages that accompany undocumented immigrant labour are factored in by large developers when securing financing, and ultimately this makes the price that tourists have to pay more palatable. Market or collective-bargaining wages in the construction phase of tourist development would greatly increase capital costs and impact the viability of the tourist enterprises. The surplus of immigrant labour available in Orlando in the 1990s drove the expansion of tourist facilities, allowing the metropolis to compete favourably with other destinations whose labour costs may have been less attractive to developers, and thus enhancing Orlando's consumptive appeal.

Construction, of course, is an essential mechanism in producing the possibility of spatial representation. The high-profile outposts of the tourist metropolis in Orlando, like Disney, Universal Studios and Sea World, are successful within a capitalist economic structure because of their ability to wed symbolic meanings to the built environment. While the aforementioned symbolic spaces are the draw of Orlando, the reality of the tourist experience extends beyond those enclaves to spaces throughout the metropolis where the outlet malls, timeshare resorts, independent and franchised hotels, gift shops, golf courses and other tourist accoutrements are situated. Many of these smaller and dispersed spaces expanded during the 1990s tourist boom, opening further opportunities for immigrant labour in the form of construction and maintenance employment.

Occasionally, this normally 'unseen' shadow element of immigrant tourist space production is propelled to the surface in the larger public consciousness. One of the more publicized incidents speaking to the precarious nature of the workplace for many shadow immigrants took place in 1998 in Celebration, Disney's residential community-cum-tourist destination. Developed as part of Disney's diversification and expansion in the 1990s, it was one the larger construction projects then underway in Orlando. In February 1998, during the early phase of construction, United States Border Patrol agents raided a construction site, arresting 16 Mexican migrants and processing them for immediate deportation.[4]

Raids such as this are not infrequent in many communities in the United States. Yet two things make this incident particularly significant. It shows how the arrest of immigrant manual labourers working in a *Disney* development can momentarily shatter the veneer of representation which is essential to the tourist space production. In addition, it highlights the position of the shadow immigrant workforce within a series of economic relationships that create a working environment of material uncertainty. Disney's Celebration has been sold, marketed and consumed using the same nostalgic cultural referents to a mythical 'Americana' that are evident in its theme parks (Bartling 2004). The exposure of shadow immigrants literally building the concrete manifestation of a mythos which largely excludes them brings to light the imperfect aspect of representation. As employees of Disney's contractors' subcontractors, they are also victims of contingent and precarious labour arrangements. High-profile companies such as Disney normally exclude shadow-sector undocumented immigrants from their regular employee rolls. Still, much of the necessary labour input for both constructing and maintaining corporate tourist space is funnelled through subcontractors that make copious use of such labour.

Immigrant production of Orlando's spaces of consumption

I have argued above that Orlando must be understood as a space where the often conflicting processes of consumption and production are mediated, and where the relations of power that inform such conflicts have specific effects on the regions' labourers. To be attractive, a commodified space must embody a 'cogent' narrative and must suspend referents to the 'real'. In the words of a Canadian columnist, who was responding to Orlando natives who complain that the tourist area of Orlando is not emblematic of the region's offerings:

> Why in the world would a family from Ontario, say, drive more than 1300 miles to look at a non-touristy Orlando.... Touristy areas are touristy because people like them. People swarm to the Orlando area for one reason: to be entertained, and to be entertained in a way they can't at home.[5]

Among the most 'entertaining' elements of Orlando's tourist industry are the large-scale theme parks. Walt Disney's vast properties are the oldest and largest in the Orlando region, making the company nearly synonymous with the city. The image of Orlando in the mind of the tourist is equated with Disney, and within US culture the company's theme parks have become the destination of a 'middle-class hajj' (Ritzer 2000). From a consumptive standpoint, Orlando has barely any identity other than being the site of Disney. As such, Disney is the single major entity for structuring the city-as-commodity.

Within Disney's properties, perhaps Epcot is best situated to reveal how the metropolis of consumption has differential effects on immigrant workers. As we have seen, the explicit production of the tourist metropolis of Orlando depends on a sizeable pool of undocumented immigrants who perform the necessary tasks of construction and maintenance behind the shadows of the spaces of play. At the same time, however, a very different but companion force of migrant and immigrant labour is equally essential in sustaining the representative aura of the tourist metropolis. If the undocumented migrants discussed above build the tourist metropolis through their hidden labour, the largely legal, Western and middle-class foreign migrants at Disney and Epcot serve as a conspicuous reference point for consumptive representation.

Far from being in the shadows of representative consciousness, this second level of immigrant labour is unique to the Orlando tourist industry and is an integral part of the production of 'authentic' representations which is the hallmark of the new city-as-informational-commodity. I will refer to the role that these workers play in metropolitan commodity construction as a 'showcase' function. Though fewer in numbers than the immigrants occupying the shadow sector, the showcase workers are accorded a decidedly public presence. At Disney's properties, this class of labourers staffs the various pavilions in Epcot's World Showcase, and to a lesser extent in Disney's Animal Kingdom – speaking the languages, delivering the food and ostensibly 'representing' the culture of their various homelands.

The showcase workers thus provide a degree of 'authenticity' to the general Epcot experience, and they are best understood as an extension of the theming that makes the park attractive. The tourists not only see a realistic rendition of the Eiffel Tower or St Mark's Square at Epcot, but they are also entertained by native musicians and dine on their 'Tournedos grillé aux cepes et sa pommes Darphin' and 'Polletto Ruspante all'Uva' served by native speakers who can pronounce the names of the dishes. In other theme parks and hotels where the theming is not as demonstrably international, showcase workers are still visible; and while their ethnicity is not as explicitly reiterated by an environment of referential codes, their appearance lends a sense of 'cosmopolitanism' that contributes to the value-added distinctiveness of the particular tourist outpost. Showcase workers at Disney's properties are often 'marked' by their ethnicity, either by actual costumes or by the ubiquitous nametags that also indicate the wearer's home country.

The common denominator linking showcase workers is their 'safeness' as representative symbols of non-US global culture. They are 'safe' because their public presence conforms to a general sanitized image of the Other. Most are from European or European-descended (such as Canadian or Australian) backgrounds and countries. Besides speaking their native languages, they are fluent in English – making their foreignness palatable to the thousands of US tourists who travel to Orlando for a healthy dose of (safe) escapism. The touchy issues of globalization's reality – imperialism, colonial legacies, poverty, cultural fragility – are silenced, as the interactions between the consumers and those providing the enabling labour for consumption are disciplined to follow a pattern of servility. The relationship largely mimics the general disposition in broad strata of US society in terms of how people understand the US relationship with the rest of the world.

The showcase workers are best described as seasonal migrants, as opposed to immigrants who intend to settle permanently in the United States. The Walt Disney World International Program (WDWIP) is designed to bring young adults from target countries to work in the theme parks in public jobs: hospitality, food service and preparation, and cultural performance. WDWIP workers are generally under contract for three to twelve months.[6]

In many ways the WDWIP can be understood as a proto-'company town' for the globalized informational economy. As indicated above, the purpose of the programme is to provide international authenticity for the globalized theming in the attraction's various efforts to represent non-United States cultures. Thus, only workers from a select number of represented countries are eligible for recruitment.[7] But 'international authenticity' in the US tourist industry does have its limits. Potential WDWIP workers are required to exhibit good conversational English, making them effective hospitality workers in the US context. Sanitized authenticity is required. Too much conversation in a non-English tongue could make the attractions' guests feel uncomfortable.

To recruit these workers, Disney-authorized agents hold interview sessions in countries of origin and select candidates for the programme. Successful applicants are given jobs in food service, concessions or other jobs associated with the 'pavilion' of their country of origin within the themed environment. Their employment is restricted to these particular pavilions by the Q-1 visa programme, which allows temporary workers to go to the United States for cultural and educational purposes. Like many non-citizen work visas, the Q-1 requires that the employer initiate the process of application. Because of the employer initiation, workers are legally bound to work for Disney.

The agreements that Q-1 workers sign with Disney also require them to live on Walt Disney World property. Accommodations for foreign workers are provided by the company in three sprawling apartment complexes, complete with swimming pools, recreational facilities and barbeque pits, and cost around $75 (about €60) per week for a shared room. The company assigns each worker a room and a roommate. According to participants I interviewed, the company tends to assign individuals to rooms with people from a different country. The housing complexes are situated in an isolated part of Disney's property, making the workers dependent on an employee shuttle to get to work. The company does not treat the housing as a subsidy to its workers and charges market rates, arguing that because they receive free transportation to work, they are 'saving' money (Verrier 2000).

While there are considerable restrictions for many of the showcase workers as a result of the employer's ability to dictate the terms of their employment, it would be unfair to characterize their experience in a negative light. Interviews and postings on a major unofficial internet forum for alumni and prospective applicants suggests that the opportunity to live in the United States and to work for Disney is such an attractive prospect that any inconveniences that might arise from the restrictive work relationship are minimized. Much of this can be explained by the particular demographic that is targeted by Disney and its recruiters when soliciting participants for the programme. Although the company closely guards reliable demographic data on Q-1 participants at Disney, it is not unreasonable to assume that most Q-1 workers are between 18 and 25 years of age. The Q-1 designation does not allow spouses or dependants to travel with the visa holder, and virtually all my interviews and the discussions relating to employee age on public internet forums[8] indicate that the workers are typically young. They look upon the ability to work at Disney as an extended vacation or a continuation of college. To be offered employment, they already must be compatible with the 'Disney culture' of politeness, friendliness and conviviality.

Alongside Disney's own aggressive training programme that educates all its employees to the organization's corporate culture, there exists a vibrant subculture generated by programme participants. Good-natured 'hazing' is a rite of passage endured by all new arrivals to the programme.

Usually, residents of the newcomers' home country engage in a ritualized welcome. Generally these events introduce new participants to a culture of revelry and fun that mark their experiences. Many alumni of the Q-1 programme have used the internet to sustain the contact they developed as co-workers and neighbours. Content on these alumni sites tends to be nostalgic and to involve retelling tales about bacchanalian debauchery and other saturnalia. The bonds developed during these times appear so strong that many groups of former workers have periodic reunions throughout Europe and elsewhere.[9]

During their time as Disney workers, they earn just above minimum wage, have their rent deducted directly from their paycheck, and perform rather menial tasks such as food preparation and service, retail stocking and selling, and routine park maintenance. Essentially the workers perform the same tasks that Disney's domestic and shadow immigrant workforce performs, and the programme also replicates other management strategies for contingency employment deployed by Disney. In addition to the high proportion of seasonal and temporary labourers hired throughout the resort complex, Disney also maintains a College Program directed towards United States students and a related internship programme.

All of these projects are successful from Disney's standpoint to the extent that they contribute to keeping labour costs low. Their success in appealing to workers is intimately related to the consumptive ethic that permeates the development of tourist cities like Orlando. For the international showcase workers, employment at Disney is often defined more as an experience of consumption than of production. When asked by a reporter why she came to work at Epcot, an employee from Brazil contended: 'I always wanted to work here because ever since I was 8, I'd come here with my family... I love this place'.[10] In justifying the relatively low wages paid by Disney to young international and college workers, Disney executive Maria Rodriguez asserted that the programme is more about building the *capacity* for future earning power. 'It's a great networking tool. The exposure to the senior-level executive in this company is a great opportunity'.[11]

These claims speak, perhaps, to the class biases of the showcase programmes. A significant proportion of the WDWIP workers come from middle- or upper-middle-class backgrounds in their home countries. Like the Brazilian worker quoted above, many have been on holiday to Disney as kids and have quixotic expectations about their experiences as Disney workers. The requirement that these showcase workers have English language skills already presupposes that they have had considerable educational opportunities. They are in a position to ignore some of the contingencies of their contracts with Disney. They may be getting a subsidy from their parents. Free of childcare responsibilities or providing for a family, and aware that their penury is only temporary, they find the terms of the Q-1 visa palatable.

This does not mean that the experiences have been altogether positive. According to Section 101(a)(15)(Q) of the US Immigration and Nationality Act (see USCIS 2005), the Q-1 visa programme must provide a foreign national with the opportunity to work in an 'international cultural exchange programme' that is accessible to the US public, explains the culture of the home country to the public, and serves 'as a vehicle to achieve the objectives of the cultural component'. Since most jobs actually performed by members of the programme involve the simple insertion of foreign workers into pedestrian food service and custodial tasks, workers have found that the reality of work fell short of what was advertised.

Interestingly, some of the more publicized incidents of worker discontent have involved workers from Kenya and South Africa who were critical of missing the promised educational opportunities. To be in compliance with federal immigration law, the company must offer educational opportunities for Q-1 workers. Workers charged that the educational opportunities were minimal in comparison to the employment responsibilities. Because of the vastness of Disney's operation and the large number of foreign workers employed throughout the resort, individual workers report directly to managers at their immediate workplace (restaurant, ride or store) and have little control over their schedules. This makes it difficult to attend educational activities if these conflict with their designated work times. The same workers were also dismayed by the lack of possibilities to work overtime, as their airfares had initially been paid by the company and were now being withheld from their earnings. In view of their substantial debts, their market-rate rents and their inability to resign or find extra work, sympathetic members of Disney's union took up collections to help them meet their obligations. Disney's response was to suggest that 'this is a living, learning and earning experience. It's not a get-rich-quick experience'.[12]

In the case of the African workers, working for Disney in Orlando was attractive largely because they could ostensibly earn and save money to support family members in Kenya and South Africa. It is significant that temporary migrants from poorer countries were the first to lodge major complaints about the conditions of showcase work. Apparently, Disney's showcase programme, which attracts certain types of public workers in an attempt to achieve authenticity of representation, can only succeed if the workers view the work as an opportunity for conspicuous consumption rather than as an avenue to economic security.

Political efficacy in the consumptive tourist city

In the light of the bifurcated nature of the immigrant/migrant worker experience in the tourist metropolis of Orlando, and given the increasing competitiveness of cities in distinguishing themselves to attract the tourist dollar, the prospects for challenging the inequitable nature of the production of consumption are slim in most quarters. The 'showcase' migrant class that provides the image of authenticity for the cosmopolitan tourist experience shares the same class biases and penchant for relating to the world in consumptive forms as the majority of visitors to Orlando's theme parks. The transitory nature of their visas, the conditions of indentured servitude imposed by their employment contracts, and the fact that most workers are young, relatively secure economically and sincerely grateful to spend a carefree year in the Disney consumptive paradise – all these factors contribute to the overwhelming success of the programme. In programmes such as the WDWIP and the College Program, the logic of the tourist metropolis is taken to its logical (and absurd) extreme: the viewing of work itself as a consumptive opportunity.

Much more political efficacy can be found at the level of 'shadow' immigrants working in the region's tourist industry. These immigrants are not hired because they can reiterate the representative themes of the industry's properties, but because they are an easily exploited workforce on grounds of their largely undocumented status and their origin in communities of extreme poverty and low economic opportunity. Like many shadow immigrant economies, the tourist economy is one of extreme contingency. Many of these predominantly Latino and Caribbean workers are actually engaged in the same tasks as their showcase counterparts, such as food service and custodial work. Unlike the showcase workers, however, shadow immigrants are not provided with affordable housing, transportation or other accoutrements that subsidize the low wages they receive for their labour. Most are hired by small firms that subcontract their work out to the tourist enterprises of large multinational corporations like Disney.

While the shadow workers do not have the same 'built-in' support networks and ritualized welcoming practices as showcase workers, there are numerous organizations and localized movements in Central Florida that have developed rather autonomously to assert immigrant interests in the region. Groups such as the Florida Farmworker Association, the Central Florida Living Wage Coalition, the Service Employees International Union and the Hotel Employees Restaurant Employees Union have entered into organizational partnerships in recent years to highlight problems of poverty, housing, health care access and other social issues. Their efforts to develop mechanisms of social dialogue and civic interaction which actively acknowledge the contribution of immigrant labour represent a fundamental challenge to the conception of the city-as-commodity. Their efforts, at this stage, are still rather modest and their modes of political action and influence are relatively undeveloped. But as the region's economy develops within the context of an ever more competitive globalized tourist market, these alternative modes of organization will be increasingly needed. They are an essential counterweight to corporate tourist enterprises that seek to maximize profits through the increasing segmentation of immigrant, migrant and non-immigrant labour.

Notes

1 The approach has notably been employed by Mayor Michael Bloomberg of New York, who, in his own self-congratulatory state-of-the-city speech, promised that 'we'll establish a Chief Marketing Officer for the city... . And we'll be taking advantage of our brand'. This is pressing, Bloomberg argued, since '... as a city, we've never taken direct, coordinated custody of our image' (Bloomberg 2003).
2 Interview with Tirso Moreno, 26 November 2002, Apopka, Florida.
3 D. Tracy, 'Mexican migrants carve path of hope to Orlando', *Orlando Sentinel*, 5 November 2003.
4 L. Savino, 'Celebration workers face deportation', *Orlando Sentinel*, 14 February 1998.
5 J. Clarke, 'The town that Mickey built', *Toronto Star*, 8 October 1988.
6 The information discussed here on the Walt Disney World International Program was garnered from interviews with participants and from various internet sources that act as clearinghouses for individuals working in (and 'alumni' of) the programme. Disney's official website (http://disney.go.com/disneycareers/wdwcareers/international/index.html [accessed 7 February 2005]) contains very little information, but an unofficial alumni-run site provides an important 'insider' perspective (http://www.wdwip.com [accessed 7 February 2005]). The website for the major European recruiting agent also provides specific information on the international programme (http://www.internationalservices.fr/ [accessed 7 February 2005]).
7 While the company has several programmes for international workers that are not geographically based, the main Cultural Representative Program is restricted to workers from Mexico, Norway, China, Germany, Japan, Italy, Great Britain, Morocco, France and Canada. Disney's safari-themed Animal Kingdom park has an African Cultural Representative Program which is open to any African applicants (apparently the geographic specificity of the theme is less important for a consuming clientele whose knowledge of the diversity of African culture, peoples and ecosystems is minimal).
8 Such as on http://www.wdwip.com.
9 For illustrative alumni websites see: http://www.geocities.com/disneythaicast/index.html; http://www.drunkadventures.co.uk/ (all accessed 7 February 2005).
10 Cited in 'Spreading the word: Millennium Village full of cultural ambassadors', *Orlando Sentinel*, 17 December 1999.
11 L. Montanez, 'Disney training program expands in Puerto Rico', *Orlando Sentinel*, 25 July 1997.
12 R. Verrier, 'Disney disappoints some foreign workers', *Orlando Sentinel*, 22 July 2001.

References

Agricultural Statistics Board (1999) *Citrus Fruits: final estimates 1992–97*, Washington DC: United States Department of Agriculture, National Agricultural Statistics Service.

Bartling, H. (2004) 'The magic kingdom syndrome: trials and tribulations in Disney's celebration', *Contemporary Justice Review*, 7(4): 375–93.

Bloomberg, M. (2003) '2003 State of the City address', available www.gothamgazette.com/article/searchlight/20030123/203/233 (accessed 7 February 2005).

Braverman, H. (1974) *Labor and Monopoly Capital: the degradation of work in the twentieth century*, New York: Monthly Review Press.

Foglesong, R. (2001) *Married to the Mouse: Walt Disney World and Orlando*, New Haven: Yale University Press.

Hodges, A. and Mulkey, W. (2003) *Regional Impacts of Florida's Agricultural and Natural Resources Industries*, Gainesville: University of Florida, Food and Resource Economics Department.

Hodges, A., Philippakos, E., Mulkey, D., Spreen, T. and Muraro R. (2001) *Economic Impact of Florida's Citrus Industry, 1999–2000*, Gainesville: University of Florida, Institute of Food and Agricultural Sciences.

Hood, G. (2002) '2002 State of the City speech', available http://web.archive.org/web/20020911163129/www.cityoforlando.net/elected/mayor/Speeches/2002/02_03_27_stateofcity.htm (accessed 18 April 2005).

Lukács, G. [1923] (1972) *History and Class Consciousness: studies in Marxist dialectics*, trans. R. Livingstone, Cambridge MA: MIT Press; originally published as *Geschichte und Klassenbewusstsein: Studien Über marxistische Dialektik*, Berlin-Halensee: Malik-Verlag.

Marx, K. [1890] (1992) *Capital: a critique of political economy*, volume 1, trans. B. Fowkes, New York: Penguin Classics; originally published as *Das Kapital: Kritik der politischen Ökonomie*, 4th edn, Hamburg: Meissner.

Ritzer, G. (2000) *The McDonaldization of Society*, 3rd edn, Thousand Oaks CA: Pine Forge Press.

USCIS (US Citizenship and Immigration Services) (2005), 'Immigration and Nationality Act', available http://uscis.gov/graphics/lawsregs/INA.htm (accessed 7 February 2005).

Verrier, R. (2000) 'Disney plans apartments for workers', *Orlando Sentinel*, 7 January.

Part III

Ethnic diversity in urban place promotion

7 Ethnic heritage tourism and global–local connections in New Orleans

Kevin Fox Gotham[1]

This chapter uses a case study of New Orleans to examine how global forces and local organizations facilitate the growth of ethnic heritage tourism. Recent years have seen the growth of a broad literature on the role of tourism in promoting the cultural exoticism of local peoples the world over (Desmond 1999; Meethan 2001; Urry 1995: 163–70, 2002). 'Ethnic heritage tourism' refers to the varieties of leisure and travel that involve the commodification of ethnic goods and cultural activities and the staging of ethnic rituals and customs for tourist consumption (Chang *et al.* 1996). In this conception, tourism interests rework and modify representations of ethnicity and culture to make them appealing to visitors. Parallelling this process of representation is the attempt by metropolitan elites to create ethnic tourist districts such as Chinatowns (Lin 1998), to redevelop urban ethnic neighbourhoods (Boyd 2000; Hoffman 2003) and to persuade immigrants to migrate to their cities to help build a multiethnic populace and cosmopolitan culture. Many scholars have analysed efforts by tourism professionals to employ selected cultural symbols and motifs to stimulate travel to various cities and facilitate the building of tourism venues (for overviews, see Alsayyad 2001; Fainstein and Judd 1999; Hannigan 1998; Judd 2003; Kearns and Philo 1993). Yet urban scholarship on tourism still lacks specificity in analysing how, and under what conditions, tourism is a force for globalizing ethnicity (by delocalizing it and disembedding it from place, for example) as well as a force for producing and reinforcing ethnic distinctiveness *in* place.

This essay embraces a constructionist view of ethnicity to explain the development of ethnic heritage tourism. A constructionist view emphasizes the socially constructed aspects of ethnicity, such as ways in which ethnic identities, grievances, ideology and local cultures are negotiated, defined and produced by individuals and groups. Central to this approach is the idea that ethnicity is mutable, contingent, a product of social ascriptions, and a reflexive process involving internal and external forces and actors. Ethnicity can be a signifier of both physical differences and differences in language, religion, region or culture (Barth 1969; Nagel 1994,

1996; for an overview, see Kivisto 2002: 13–42). Nagel (2000: 111) conceptualizes ethnicity as:

> *performed* – where individuals and groups engage in ethnic 'presentations of self'; and *performative* – where ethnic boundaries are constituted by day-to-day affirmations, reinforcements and enactments of ethnic differences. Ethnicity is thus dramaturgical, situational, changeable and emergent.

Ethnicity can also be instrumental and strategic. It can be strategically deployed by different groups for instrumental purposes – such as to confront the values, categories and practices of the dominant culture, to challenge the dominant culture's perception of a minority group, or to contest public policy within the political arena. Wood's (1998) analysis of ethnic groups in Asia and the Pacific Rim draws attention to a development of 'touristic ethnic cultures', in which ethnic group interaction with tourism is an integral part of the construction and reproduction of the ethnic identity itself (see also van den Berghe and Keys 1984; Wood and Picard 1997). Local meanings of ethnicity shape, and are shaped by, the advertising and promotional strategies of tourism. In this way, the tourism creates new bases of struggles and conflict over meanings of ethnicity.

In this chapter I investigate the ways in which ethnicity is taking on an expanded role in urban economies, and how it is becoming a fundamental theme that distinguishes tourist destinations and entertainment-enhanced developments. In New Orleans and elsewhere, we now witness an explosion of ethnicity, as city officials and economic elites try to 'sell' their locality by harnessing its ethnic attributes – actual or perceived. Today, place promotion involves the marketing of ethnic diversity and multiculturalism using an array of symbolic devices, advertising campaigns and the like. Ethnicity is becoming more globalized *and* more localized (heterogeneous and differentiated). As I attempt to show here, the impact of global flows is evident in changes in the relations between public and private tourism institutions that link New Orleans to the global economy and to globalized ethnic tourism. The impact of local forces is evident in how tourism organizations try to simulate ethnicity and culture, using general and specific themes, to entice people to visit and spend money in New Orleans. Finally, I explore the conflicts and the struggle between different interests over the issue of *which* particular cultures and ethnic groups should be represented in tourist advertising and how they should be represented. I highlight the problems of local tourism marketing efforts as they attempt to advertise the 'diversity' and 'multiculturalism' of New Orleans, precisely when the city has become more racially *homogeneous* over the last three decades.[2]

Development of tourism in New Orleans

The 1940s and 1950s witnessed the beginning of a fledgling tourism industry in New Orleans. During the immediate post-World War II years, city officials and elites began devising strategies to increase tourist travel to the city, in the hope of improving the economic prosperity and fiscal status of the central city. The 1950s saw an impressive growth in the number of tourists and conventions coming to New Orleans. Total convention delegate spending rose from approximately $3 million in 1951 to $5.8 million in 1955 and $11 million in 1958. Dwindling urban population, burgeoning suburban development and deindustrialization from the 1960s onwards raised the spectre of economic stagnation and created the context for city leaders to accelerate the development of tourism in the city.

Deindustrialization took its toll on the manufacturing and port industries that were the backbone of the local economy. From 1967 to 1977, manufacturing jobs in New Orleans declined in every year except one. By 1977, only 11 per cent of the labour force was still employed in manufacturing, ranking the city among the lowest in the nation in terms of industrial employment (Smith and Keller 1986). The specialized nature of the region – with tight-knit linkages between the larger oil and manufacturing firms and the smaller companies that supplied them with materials – meant that the metropolitan area was more vulnerable to deindustrialization than a diversified economy might have been. The oil market crash from 1982 to 1987 depressed the local jobs market still further, triggering a dramatic increase in housing foreclosures and the outmigration of thousands of middle-class families from the city and the whole metropolitan area (Lauria and Baxter 1999). The effects of global restructuring within the chemical and oil industries were especially pernicious. The urban population plummeted from a high of 627,525 in 1960 to an all-time low of 484,674 in 2000. The city lost more than 34,000 residents during the 1960s, over 35,000 in the 1970s, over 60,000 in the 1980s and about 38,000 in the 1990s.

Since the 1980s, civic and business leaders in New Orleans have forged close institutional links and developed several public–private partnerships to pursue tourism as a means of stimulating inward investment and urban revitalization (Gotham 2002). The various components of this tourism strategy have included the building of a domed stadium, a festival mall, a massive convention centre, a land-based casino, a theme park and a World War II museum, as well as the staging of mega-events like the 1984 World's Fair, periodic Super Bowls and (Nokia) Sugar Bowls, and NCAA basketball tournaments. According to municipal data, the city attracted over 13 million visitors in 1998, an increase of 63 per cent since 1990. The hotel industry has also prospered in recent decades. The number of hotel rooms increased from 4750 in 1960 to 10,686 in 1975 and to 19,500 in 1985. In 1990, the metropolitan area had about 25,500 hotel or motel rooms, a figure that grew further to 28,000 in 1999 and almost 34,000 by 2000. The convention market has expanded immensely, too, since the 1970s. The city hosted 764 conventions in 1976, 1246 in 1987, 2000 in 1992 and 3261 in 1999. Although the numbers declined following the national economic downturn and the 11 September 2001 tragedy, the city and state governments approved financing in 2003 to enlarge the Ernest Morial Convention Center. The expansion will make it the fourth largest convention centre in the USA, and it is to attract more than 1.4 million visitors to the city (City of New Orleans 2000). Tourism jobs number over 31,000, now making up nearly 30 per cent of the metropolitan area's total employment figure. The economic significance of tourism is put at more than $4 billion (around €3 billion) annually.

The 1990s were the beginning of a new era of intensified tourism development and place marketing in New Orleans. Four key developments occurred. First, in 1990, the State of Louisiana and the City of New Orleans established the New Orleans Tourism Marketing Corporation (NOTMC), a private, non-profit economic development agency designed to foster job growth and economic revitalization by marketing New Orleans as a leisure destination. Through a programme of advertising and public relations, the NOTMC tries to boost hotel occupancy when tourism is slow, specifically during the summer and the weeks between Thanksgiving and New Year's Day. The organization is funded by a hotel occupancy privilege tax, a Regional Transportation Authority (RTA) tax, the City of New Orleans General Fund, and lease agreements by Harrah's Casino with the municipal authority and the hotel industry. Second, also in 1990, thirteen African American business owners established the Black Tourism Network (BTN) to expand opportunities for African Americans in the tourism industry. In 1999, the BTN was renamed the New Orleans Multicultural Tourism Network (NOMTN) and extended its mission to promote the cultural diversity of the city. Third, in the mid-1990s the City of New Orleans established the Mayor's Office of Tourism and Arts to liaise between the tourism industry and arts organizations. Finally, in 1995 the State of Louisiana and the City of New Orleans adopted statutes that earmarked part of the local hotel–motel tax for the New Orleans Metropolitan and Convention and Visitors Bureau (NOMCVB) with the aim of boosting foreign travel to the city and promoting the city internationally as a leisure and convention destination. All four of these organizations have become central forces in the development of the New Orleans tourism industry, and they play strategic roles in broadening the base of travellers that visit each year.

Public–private relations and the role of transnational corporations

Recent years have seen major changes in the production of local ethnic and cultural festivals as a result of the expansion of the New Orleans tourism industry. One major transformation has been the growth of new public–private networks linking transnational corporations with local organizations and agencies. Today, transnational corporations are the chief sponsors of local festivals such as the Southern Decadence celebration, the French Quarter Festival, the Jazz and Heritage Festival, the Essence Music Festival, the Satchmo Festival and the Swamp Fest. The McDonald's corporation, Pepsi Cola and Southwest Airlines all underwrite the Swamp Fest, held every October. Over the past eight years, the Essence Music Festival has grown from a predominantly African-American festival into a broad-based multicultural celebration attracting 223,000 people for this three-day event held every July. In 1996, Coca Cola became a sponsor of the Essence Festival and increased its contribution in 1998 to become the title sponsor. The festival's marketing reach has been extended in recent years, as it targeted international consumers under a partnership that Essence Communications established with AOL Time Warner in 2000. This relationship began with online promotions in 2001, and it grew in 2002 with advertisements in magazines such as *Time*, *People* and *Entertainment Weekly*.

For decades, city residents and leaders had been accessing localized networks and organizations to stage festivals that were primarily aimed at local residents and tourists. Festivals and other money-making spectator events were supplementary to larger economic development initiatives designed to attract manufacturing industries and investment. Today, in contrast, local political officials and economic elites have come to rely on extralocal networks and organizations, including transnational marketing firms and advertising agencies, to produce and promote the festivals. Festivals are no longer ancillary to manufacturing and heavy industry, but are now seen as a centrepiece of local pro-growth strategies to attract domestic and international tourists to spend money in the city. In addition, transnational corporate advertising, promotions and other attempts to fuel consumption now dominate the planning and design of all the ethnic and cultural festivals in New Orleans.

This increasing 'corporatization' of local ethnic events and festivals translates into cultural standardization as well as differentiation. On the one hand, the production of festivals becomes more homogenized, in the sense that transnational tourism-oriented firms (hotels, entertainment companies, destination marketing firms and so on) embrace similar marketing strategies and advertising campaigns. On the other hand, ethnic and cultural festivals in the city and region become more differentiated, specialized and heterogeneous, as local groups and organizations construct and produce different meanings of the 'local' to persuade people to visit their towns and festivals, to buy local products, and to view and experience local cultures and sites. As cities increasingly compete for tourist dollars and investment, the pressure to produce differentiated products for targeted markets – to mine culture and place for profit – gives rise to even more spectacular ethnic images and ethnic tourist sites.

In New Orleans, processes of globalization and localization are occurring simultaneously. Each year the city hosts dozens of annual events – ranging from art and craft festivals highlighting Creole, Cajun and other cultures to ethnic festivals celebrating Hispanic, Irish, German, French, African and Caribbean heritages. Local festivals serve important 'local' goals, such as drawing tourists and generating profits for local businesses, as well as the 'global' goal of increasing public exposure to extralocal corporate advertising. In this sense, the marketing of local ethnic images is part of what scholars call the 'globalization of the local', whereby local symbols, motifs and images are advertised to a global audience (Chang 2000a, 2000b). Of course the consumption of ethnic goods, artefacts and so on has always occupied a central place in ethnic heritage tourism. Along the way, tourists have been motivated to buy ethnic products, trinkets and gifts to signify their attendance at a local festival or tourist attraction. Yet today the consumption of goods is increasingly the *paramount* objective of ethnic heritage tourism. In this new orientation, local festivals, celebrations and other tourist attractions can become a means to sell other commodities such as ethnic souvenirs or ethnic art, whereas visits to New Orleans also fuel a broader interest in New Orleans food, movies, books, music, art, architecture, ethnic groups and culture. Tourist advertising focuses and particularizes the destination image by auditing and developing local resources, promoting a sense of place, and thereby activating a desire to visit the city. Seen in this light, the production of ethnic festivals is now a globalized process connecting exogenous forces like multinational corporations and capital flows to the locally based actions of residents, elites and consumers.

Niche marketing, branding and the 'holy trinity' of New Orleans tourism

A second major transformation in the production of local ethnic festivals is visible in promotional strategies, which are now less targeted to mass markets and more geared to creating and exploiting 'niche markets' using sophisticated branding techniques. In tourism marketing, ethnic groups are targeted as niche markets. Examples include groups of African-Americans, Asians, Hispanics and various European nationalities. *Niche tourism* refers to the development of new forms of cultural fragmentation and spatial differentiation that divide tourists, tourist markets and spaces of tourist consumption into ever-smaller segments or niches, thus increasing heterogeneity (Meethan 2001: 72–3). On the one hand, niche markets develop out of local efforts to stimulate socioeconomic development, to bolster the local tax base and to attract visitors. On the other hand, they are a response to globalizing tendencies like the growth of international differences, the proliferation of new information and communication technologies, the flow of immigrants and the emergence of new ethnic communities in cities (Ioannides and Debbage 1997; Wonders and Michalowski 2001). The logic of niche marketing implies the pursuit of ethnic and cultural 'branding' – the act of marking something with a distinctive name to identify a product or manufacturer (Greenberg 2000). Discussing their aggressive marketing of African-American consumers and the Essence Festival, Sandy Shilstone, executive vice president of the New Orleans Tourism Marketing Corporation, indeed stated graphically: 'We want to make sure that the New Orleans name is branded'.[3]

In their discussions of place marketing and branding, Molotch (2002) and Greenberg (2000, 2003) have argued that urban redevelopment is as much about 'accumulating brand value' as it is about zoning, code enforcement and the other more conventional aspects of planning and policy. Hannigan's (1998, 2003) discussion of the entertainment-based redevelopment strategies of 'fantasy cities' points out that branding requires the development of promotional 'synergies' with the arts, entertainment and tourism. In New Orleans, branding involves the differentiation of places in the city and the valorization of multiculturalism and diversity – the intent being to enhance cultural capital and generate consumer interest in formerly unattractive places. One tourism agent from the New Orleans Multicultural Tourism Network told me about their niche marketing strategy:

> The New Orleans Multicultural Tourism Network promotes the diversity of New Orleans through the collateral pieces that we produce. We show and promote New Orleans as a place that African Americans, Hispanics and Latinos can enjoy and want to visit. If you want to go to a restaurant, we have black restaurant owners. You want to go on an African American history tour centred around the city? Well, we have those things and we have people that can provide those, and we market them to people. We have Hispanic populations, we have Asian populations, we can point people in the right direction for those things. We promote the Isleños, the Germans – all the folks that made New Orleans what it is.

Brand marketing campaigns and image-building involve a mix of claims to distinction, on the one hand, with assurances of predictability and comfort based on homogeneity and standardization. In New Orleans, the narrative of distinction is constructed around three themes – history, music and food. These constitute the 'holy trinity' of New Orleans tourism. The quote below from one official of the New Orleans Tourism Marketing Corporation (NOTMC) highlights this vital triad that connects and unites the disparate elements of the city and region.

> The images the New Orleans Tourism Marketing Corporation uses to promote tourism are evocative, and they emphasize the holy trinity of

> New Orleans tourism: food, music and history. All the cultures of
> New Orleans emphasize these elements, and we use them to promote
> the city and its peoples. We advertise in regional markets, consumer
> magazines and on cable television. Our emphasis is on the authentic-
> ity and the heritage of New Orleans.

Food, history and music are metaphors of distinction that represent the
lifestyles and cultural values of different ethnic groups. What foods
people choose, how they prepare and eat them, and what foods and drinks
they abstain from express different cultural beliefs and values, and they
thus become markers of New Orleans culture. In *The Elementary Forms
of Religious Life*, Emile Durkheim ([1912] 1995) drew attention to the
role that 'collective representations' – cultural beliefs, moral values and
ideas – play in creating a symbolic world of meanings within which a
cultural group lives. Durkheim's theory of culture starts from the claim
that the major symbolic components of culture are *representations* which
are *collectively* produced, reproduced, transmitted and transformed.
Different groups create unique kinds of music and dance, play different
musical instruments and prefer listening to particular styles of music
during their cultural celebrations and rituals. Through their cuisine and
music, people maintain and reproduce collective identity and create new
cultural bonds – the 'invention of tradition' in Eric Hobsbawm's (1983)
famous formulation. Advertising New Orleans as a site of delicious food,
quality music and rich history fulfils the vital goals of showcasing the
different local cultures, attracting diverse kinds of tourists and generating
business opportunities within the local tourism industry. This narrative's
lack of specificity leaves it open to many interpretations, thus creating a
discursive space for the inclusion of different peoples and ethnic backgrounds. For example, the website of the New Orleans Multicultural
Tourism Network (2003) tells us that 'New Orleans is a gumbo of cultures
that blend together, yet maintain their own unique flavor'. According to
one official of the New Orleans Metropolitan Convention and Visitors
Bureau (NOMCVB):

> Each and every niche market we attack a little differently. And the
> one thing we constantly say is that the reason you'll enjoy your stay
> in New Orleans is because New Orleans plans its environment around
> its people. We love good food, we love music and entertainment.
> Every festival that we put on, we do not put on for the visitor – we put
> it on for ourselves. And we talk about our history, our culture, all
> the different arts. And each visitor finds something out of those that
> they want.

One person involved with the New Orleans Museum of Art (NOMA)
told me about how the inclusion of people of colour in tourism marketing

helps to raise public awareness of the contributions of diverse groups of people to the city's development, thus expanding meanings of the local and opening up new avenues of reinterpretation:

> It is said that music, food and architecture are the three main draws to New Orleans. Then it is very important that people understand the contributions of people of colour to that culture. What we have tried to do is to show the contribution of African American builders, architects and designers in the development of our historic neighbourhoods, whose ancestors still survive and still ply their trade in the city! Now this material is finally being incorporated in the narratives of tours, and I think it opens up our tourism product and exposes us, our city, to a reinterpretation that broadens our market and brings greater public awareness and exposure. Because now you're giving people a much richer overview of what the city is all about than what they got in the past.

In the above two quotes, the general themes of food, music and history are employed as signs of uniqueness that express ethnic diversity but do not signify cultural assimilation or loss of distinctiveness. The reference to these three themes directs attention to a 'rich past' of ethnic diversity and heritage that feeds into and supports a 'dynamic present' of cultural promotion led by the tourism industry. The elements of the holy trinity of New Orleans tourism – food, music and history – connect to one another by representing pleasurable experiences (eating, listening to music and gazing upon historic buildings, art and artefacts) as *consumption-based* entertainment activities. Local tourism groups seek to forge emotional linkages between the signifier 'New Orleans' and the potential consumers – including residents, investors and tourists – in such a way that just hearing the name of the city arouses a whole series of pleasurable images and sentiments, thus kindling a desire to 'experience' the city. Urry (1995: 132) suggests that tourism is about the accumulation of exotic 'experiences' which are the anticipated outcomes of the 'tourist gaze', where places are chosen to be gazed upon because they arouse a stronger anticipation of pleasure than what one would normally encounter in everyday life. According to Britton (1991: 465), 'tourists are the "armies of semiotics" for whom the identification and collection of signs are "proof" that experiences have been realized'. In this respect, the marketing of ethnic 'experiences' becomes a lucrative tourism strategy that includes tie-ins with the buying and selling of other New Orleans products (like art and souvenirs). The various actors and organized interests that market New Orleans give people a choice of goods and services to consume. Yet what they seek to limit, if not eliminate (to quote Ritzer and Liska 1997: 143, emphasis in original), 'is our ability *not* to consume'.

Ethnicity for whom? Image and reality of ethnic heritage tourism

Marketing cities as places of ethnic diversity does not necessarily encourage or translate into multiculturalism or ethnic plurality. While tourism advertisements proclaim New Orleans as a multicultural city, it is actually home to only two major racial groups – African-Americans and whites – who constitute over 90 per cent of the urban and suburban population. According to US Census Bureau data, in the year 2000 the city of New Orleans proper had a total population of 510,369, with whites comprising 29.3 per cent, blacks 64.5 per cent, Hispanics 3.0 per cent and Asians 2.5 per cent (Table 7.1). The irony is that, even as the tourism industry was intensifying its efforts to extol and market the diversity of the city and region, New Orleans was becoming steadily *less* racially and ethnically heterogeneous as a result of the ever-growing share of blacks in the population. In 1960, blacks made up 37.2 per cent of the city's population and whites 62.6 per cent; in 1990 these figures had reversed to 59.0 and 35.5 per cent; and as of 2003, blacks constituted 67.3 per cent of the city population compared to 28.1 per cent whites. Blacks now make up a higher percentage of the population in Orleans Parish (the city proper) than they do in any other parish or county of over 100,000 people in the entire USA.[4]

In short, New Orleans's racial and ethnic composition does not reflect a multicultural population, despite the promotional images of ethnic diversity promulgated by local tourism organizations. In fact, New Orleans contains one of the *smallest* immigrant concentrations among the country's major metropolitan areas, at only 6 per cent. The 2000 census recorded 64,169 foreign-born residents in the New Orleans metropolitan area, a mere 4.8 per cent share of the total population. Of them, a greater percentage lived in the suburbs than in the central city (Tables 7.2 and 7.3). Although the city has historically indeed been a gateway city for immigrants, local public officials have acknowledged the paucity of immigrant enclaves in their city, a loss of immigrants to Miami, California and other areas, and a lack of a cultural base and amenities to attract Asian and Hispanic tourists to the city. Data from the Lewis Mumford Center for Comparative Urban and Regional Research (2003) shows that New Orleans's degree of attractiveness to immigrants declined during the 1990s. Table 7.4 compares New Orleans to all 331 metropolitan areas in the United States in terms of newly arrived immigrants in 1990 and 2000.

The incongruity between the image of cultural heterogeneity and the reality of racial homogeneity has not been lost on local residents, who have decried the lack of ethnic diversity in the city, and in particular the low numbers of Asians living in New Orleans. According to Lucy Chun, leader of the New Orleans Asian/Pacific American Society, 'There is no culture for … Asians in New Orleans. Life for Asians here can be dull and

Table 7.1 Racial and ethnic composition of New Orleans population, 1990–2000

	1990		2000	
	Total	*% of total population*	*Total*	*% of total population*
Total population				
Metropolitan area	1,285,270		1,337,726	
Central city	521,062		510,369	
Suburbs	764,208		827,464	
Non-Hispanic white				
Metropolitan area	762,858	59.4	731,514	54.7
Central city	185,205	35.5	149,742	29.3
Suburbs	577,653	75.6	581,772	70.3
Non-Hispanic black				
Metropolitan area	442,939	34.5	503,108	37.6
Central city	307,643	59.0	329,241	64.5
Suburbs	135,296	17.7	173,867	21.0
Hispanic				
Metropolitan area	53,923	4.2	58,545	4.4
Central city	17,849	3.4	15,513	3.0
Suburbs	36,074	4.7	43,032	5.2
Asian				
Metropolitan area	21,917	1.7	33,220	2.5
Central city	9,825	1.9	12,645	2.5
Suburbs	12,092	1.6	20,575	2.5

Source: Lewis Mumford Center for Comparative Urban and Regional Research (2003)

Notes: *Metropolitan area* includes Jefferson Parish, Orleans Parish, Plaquemines Parish, St Bernard Parish, St Charles Parish, St James Parish, St John the Baptist Parish and St Tammany Parish.
Central city includes Orleans Parish (boundaries coterminous with the City of New Orleans).
Suburbs include all parishes outside of Orleans Parish.

boring. They have no reason to come here, or stay. Culture is important'.[5] As another person told me,

> In the last 20 years there have been few Asian people locating to live in New Orleans.... While there is a lot of history and culture to New Orleans, and Asians want to come to see this, there is no real, relatively permanent Asian presence in New Orleans to keep Asians here. Lots of Asians want to come to the city to see the Mardi Gras, [but] there are few Asian restaurants and there is virtually nothing here in New Orleans to pull Asians here to live.

Table 7.2 Immigrant admissions to New Orleans, 1991–1998

Country	Number of immigrants
Vietnam	2587
Honduras	2175
China*	841
India	630
Philippines	573
United Kingdom	347
Guatemala	319
Mexico	309
Cuba**	292
Jordan**	245

Source: Federation for American Immigration Reform (2003)

Note: The Center for Immigration Studies reported in October 2001 that 13,316 legal immigrants had indicated an intention to settle in the New Orleans metropolitan area between the 1991 and 1998 fiscal years. This number did not include persons granted legal immigrant status as a result of the 1986 amnesty for illegal aliens.
* Partial data, includes Hong Kong and Taiwan.
** Partial data.

With regard to other ethnic groups, the issue is not so much exclusion as a false or distorted representation of them in the tourism advertising. As one historic preservationist put it,

> It is true that tourism commercializes our past and culture. For example, it gets confusing for people when they try to figure out the difference between Creole and Cajun. A lot people confuse these all the time and that's because you see them advertised all over the city and in the state. Any food is supposed to be better-tasting if it is Creole or Cajun or both. Just the label is supposed to express the quality of the food. Local people know the difference, but these two great cultures get conflated in tourism advertising, and tourists don't know what they mean except some sort of feeling that they are good and signify quality. We try to make sure that the tourism industry advertises our culture accurately.

As cities compete to distinguish themselves on a global stage, they try to accumulate marks of distinction to authenticate their claims to uniqueness that yield tourism investment. The term 'ethnicity' has a great deal of symbolic value and utility for cities, at least for North American cities. Its use does not rule out any particular group, and it can be used to refer to almost any activity performed by various cultural performers. Ideologically, the term 'ethnicity' is not actively resisted by consumers

Table 7.3 US-born and foreign-born residents of New Orleans, 1990–2000

	Metropolitan area		Central city		Suburbs	
	1990	2000	1990	2000	1990	2000
Total population	1,285,270	1,337,726	521,062	510,369	764,208	827,464
US-born	1,232,265	1,273,557	499,723	489,118	732,542	784,439
% of total	95.9	95.2	95.9	95.8	95.9	94.8
Foreign-born	63,005	64,169	21,339	21,144	31,666	43,025
% of total	4.9	4.8	4.1	4.1	4.1	5.2
Immigrated in past decade	20,008	20,531	8,470	7,027	11,538	13,504
% of total	1.6	1.5	1.6	1.4	1.5	1.6
Immigrated earlier	32,997	43,638	12,869	14,117	20,128	29,521
% of total	2.6	3.3	2.5	2.8	2.6	3.6

Source: Lewis Mumford Center for Comparative Urban and Regional Research (2003)

Table 7.4 US national rankings of New Orleans relative to 331 metropolitan areas in terms of attractiveness to immigrants, 1990–2000.

New Orleans	1990	2000
Metropolitan area	55th	86th
Central city	59th	105th
Suburbs	45th	64th

Source: Lewis Mumford Center for Comparative Urban and Regional Research (2003)

and tourists, nor does it carry the negative connotation that the term 'race' has for many people. Indeed, the proliferation of ethnicity and the absence of race in tourism discourse suggests a political logic to the practice of place marketing. Reflecting Gottdiener (1997, 2000), who analyses the production of theming, tourism marketing campaigns that use ethnic imagery, symbols and themes are strategic and methodical. They are designed to enhance the power and status of tourism marketers as *community experts* who impart social values and construct 'good places' to visit. These marketing campaigns also attempt to silence alternative readings of the New Orleans cultural landscape. What the local tourism industry is seeking to promote is an ersatz ethnicity of no offence.[6] Tourism marketing promotes ethnic distinctiveness using particularity, diversity, ambiguity or repetition to appeal to fragmented niche market segments. Undefined, yet symbolically powerful, signs of ethnicity tap into and exploit consumer desires to travel to different places and see different groups. Marketing ethnicity is both a political project and a discursive device. As a political project, it represents ethnic and cultural difference as unity or identity. As a discursive device, it tries to legitimate the creation and re-creation of new cultures and ethnic groups. In a globalized world of rapid flows of people, money and images, local culture and ethnicity are increasingly produced for visitors, and ethnic identities, in their turn, are constructed out of the very images created for tourists.

Conclusion

Tourism development in New Orleans is a dynamic process which involves not only economic and social relations extant at the local level, where particular conflicts and developments occur, but also relations and processes that are global in scale and highly complex in character. Conceptualizing tourism as a product of both local and global forces helps us understand how *places* continue to retain their distinctiveness in an era of unprecedented global flows and novel patterns of global interconnectedness. New Orleans today remains one of the world's top urban tourism

destinations, and local leaders court large transnational hotel and entertainment firms to invest in the city. Yet the development of tourism and
the promotion of culture and ethnicity are firmly undergirded by, and tied
in with, local distinctions. Globalization is usually depicted as an exogenous and macro-level force that exists a priori, 'out there', and which is
relatively inaccessible to local people. Yet the global does not eradicate
the local, because particular cultural symbols, actions, power relations and
other forces from 'within' play a part in the mediating and channelling of
global forces. Viewing ethnicity as constructed, fluid and changing
can sensitize us to the role that tourism plays in differentiating and prolif-
erating signs and symbols of ethnicity through the development of niche
markets. The development of ethnic heritage tourism in New Orleans is
a three-way process involving extralocal forces (transnational corporations and tourists), local people and organizations (entrepreneurs and
ethnic groups), and tourism organizations that mediate the global–local
relationship.

This chapter has drawn attention to the multifaceted ways in which
city officials and tourism organizations use ethnicity in strategic ways
to attract tourists, to engineer economic development and to control
people and spaces. As a set of architectural themes, ethnicity now plays a
central role in shaping urban redevelopment strategies based on historic
preservation and the marketing of local heritage. As a set of advertising
images and slogans, ethnicity reflects the interests of tourism officials and
boosters in adapting and remoulding ethnic images of New Orleans to
make it desirable to the targeted consumer. Through the medium of
place promotion and advertising, tourism organizations transform ethnicity and ethnic identity into a sign or a brand image, a saleable item for
market exchange. MacCannell (1973, 1992: 168) suggests that tourism
systematically purveys a 'staged' or 'reconstructed' ethnicity – a rhetori-
cal, symbolic expression of cultural difference that is commodified,
packaged and sold to tourists. Whether ethnic differences are 'real' or
manufactured is not significant, because tourism is 'the reading of touristic signs emptied of all meaning except the signification of difference'
(Goss 1993: 686). Which ethnic identities are marketed, how they are
marketed, and who is targeted by the promotional images reflects a
highly selective reality which is governed by profit considerations. These
points demonstrate why it is essential to understand how the choices
and decisions surrounding the production of ethnic heritage tourism and
place-building are structured by underlying inequalities of wealth and
power. Although ethnic heritage tourism expands the repertoire of place
promotion, it does not necessarily give local people control over the deci-
sion-making apparatus in tourism politics. Certain groups and individuals
are more likely to benefit than others. City marketing images and strategies are created by powerful interests, often in ways that allow no real
alternatives.

Notes

1 I wish to thank Jan Rath, Joane Nagel, Peter Kivisto, April Brayfield and participants in the Tulane University Sociology Colloquium (31 October 2003) for comments on a previous draft. I also greatly acknowledge the comments offered by the participants in Immigrant Tourist Industry, a European Science Foundation (ESF) Exploratory Workshop on the Commodification of Cultural Resources in Cosmopolitan Cities, held at the University of Amsterdam, Institute for Migration and Ethnic Studies (IMES), Amsterdam, the Netherlands, 7–9 December 2003.

2 I use a combination of primary and secondary data in this paper. The secondary data come from government documents, planning reports and newspaper articles. The primary data come from in-depth, semi-structured interviews with 30 local residents who had first-hand knowledge and experience with the development of the tourism industry in New Orleans. I recruited these informants through snowball sampling.

3 'The essence of the deal', *New Orleans Times Picayune*, 29 October 2002.

4 M. Schleifstein, 'N.O. has highest ratio of black citizens', *New Orleans Times Picayune*, 19 September 2003.

5 'East Asian market planned', *New Orleans Times Picayune*, 31 July 2002.

6 I thank Peter Kivisto for suggesting this point to me.

References

Alsayyad, N. (2001) 'Global norms and urban forms in the age of tourism: manufacturing heritage, consuming tradition', in N. Alsayyad (ed.) *Consuming Tradition, Manufacturing Heritage: global norms and urban forms in the age of tourism*, London: Routledge.

Barth, F. (ed.) (1969) *Ethnic Groups and Boundaries: the social organization of culture difference*, Boston: Little, Brown.

Boyd, M. (2000) 'Reconstructing Bronzeville: racial nostalgia and neighborhood redevelopment', *Journal of Urban Affairs*, 22(2): 107–22.

Britton, S. (1991) 'Tourism, capital and place: towards a critical geography of tourism', *Environment and Planning D: Society and Space,* 9(4): 451–78.

Chang, T.C. (2000a) 'Renaissance revisited: Singapore as a "global city for the arts"', *International Journal of Urban and Regional Research*, 24(4): 818–31.

Chang, T.C. (2000b) 'Singapore's Little India: a tourist attraction as a contested landscape', *Urban Studies*, 37(2): 343–66.

Chang, T.C., Milne, S., Fallon, D. and Pohlman, C. (1996) 'Urban heritage tourism: the global–local nexus', *Annals of Tourism Research*, 23(2): 284–305.

City of New Orleans (2000) *Master Plan Issues Paper*, New Orleans: City of New Orleans.

Desmond, J. (1999) *Staging Tourism*, Chicago: University of Chicago Press.

Durkheim, E. [1912] (1995) *Elementary Forms of Religious Life*, trans. K.E. Fields, New York: Free Press; originally published as *Les formes élémentaires de la vie réligieuse*, Paris: Alcan.

Fainstein, S.S. and Judd, D.R. (1999) 'Global forces, local strategies and urban tourism', in D.R. Judd and S.S. Fainstein (eds) *The Tourist City*, New Haven: Yale University Press.

Federation for American Immigration Reform (2003) 'Metro Area Factsheet: New Orleans, Louisiana MSA', available http://www.fairus.org/ (accessed 25 January 2005).

Goss, J.D. (1993) 'Placing the market and marketing place: tourist advertising of the Hawaiian islands, 1972–92', *Environment and Planning D: Society and Space*. 11(6): 663–88.

Gotham, K. (2002) 'Marketing Mardi Gras: commodification, spectacle and the political economy of tourism in New Orleans', *Urban Studies*, 39(10): 1735–56.

Gottdiener, M. (1997) *Theming of America: dreams, visions and commercial spaces*, Boulder CO: Westview.

Gottdiener, M. (ed.) (2000) *New Forms of Consumption: consumers, culture and commodification*, Lanham MD: Rowman and Littlefield.

Greenberg, M. (2000) 'A social history of the Urban Lifestyle Magazine', *Urban Affairs Review*, 36(2): 228–63.

Greenberg, M. (2003) 'The limits of branding: the World Trade Center, fiscal crisis and the marketing of recovery', *International Journal of Urban and Regional Research*, 27(2): 386–416.

Hannigan, J. (1998) *Fantasy City: pleasure and profit in the postmodern metropolis*, New York: Routledge.

Hannigan, J. (2003) 'Symposium on branding, the entertainment economy and urban place building: introduction', *International Journal of Urban and Regional Research*, 27(2): 352–60.

Hobsbawm, E. (1983) 'Introduction: inventing traditions', in E. Hobsbawm and T. Ranger (eds) *The Invention of Tradition*, Cambridge: Cambridge University Press.

Hoffmann, L.M. (2003) 'The marketing of diversity in the inner city: tourism and regulation in Harlem', *International Journal of Urban and Regional Research*, 27(2): 286–99.

Ioannides, D. and Debbage, K. (1997) 'Post-Fordism and flexibility: the travel industry polyglot', *Tourism Management*, 18(4): 229–41.

Judd, D.R. (ed.) (2003) *The Infrastructure of Play: building the tourist city*, Armonk, New York: M.E. Sharp.

Kearns, G. and Philo, C. (eds) (1993) *Selling Places: the city as cultural capital past and present*, Oxford: Pergamon.

Kivisto, P. (2002) *Multiculturalism in a Global Society*, New York: Blackwell.

Lauria, M. and Baxter, V. (1999) 'Residential mortgage foreclosure and racial transition in New Orleans', *Urban Affairs Review*, 34(6): 757–86.

Lewis Mumford Center for Comparative Urban and Regional Research (2003) 'Diversity in black and white', University at Albany, State University of New York. Available http://mumford1.dyndns.org/cen2000/BlackWhite/BlackWhite.htm (accessed 29 November 2003).

Lin, J. (1998) *Reconstructing Chinatown: ethnic enclave, global change*, Minneapolis: University of Minnesota Press.

MacCannell, D. (1973) 'Staged authenticity: arrangements of social space in tourist settings', *American Journal of Sociology*, 79(3): 589–603.

MacCannell, D. (1992) *Empty Meeting Grounds: the tourist papers*, London: Routledge.

Meethan, K. (2001) *Tourism in Global Society: place, culture, consumption*, New York: Palgrave.

Molotch, H. (2002) 'Place in product', *International Journal of Urban and Regional Research*, 26(4): 665–88.

Nagel, J. (1994) 'Constructing ethnicity: creating and recreating ethnic identity and culture', *Social Problems*, 41(1): 152–76.

Nagel, J. (1996) *American Indian Ethnic Renewal: Red Power and the resurgence of identity and culture*, New York: Oxford University Press.

Nagel, J. (2000) 'Ethnicity and sexuality', *Annual Review of Sociology*, 26(1): 107–33.

New Orleans Multicultural Tourism Network (2003) 'Soul of New Orleans'. Available http://www.soulofneworleans.com (accessed 21 January 2003).

Ritzer, G. and Liska, A. (1997) '"McDisneyization" and "post-tourism": contemporary perspectives on contemporary tourism', in C. Rojek and J. Urry (eds) *Touring Cultures: transformations in travel and leisure*, London: Routledge.

Smith, M.P. and Keller, M. (1986) '"Managed growth" and the politics of uneven development in New Orleans', in S. Fainstein, N.I. Fainstein, R.C. Hill, D.R. Judd and M.P. Smith (eds) *Restructuring the City: the political economy of urban redevelopment*, M.P. New York: Longman.

Urry, J. (1995) *Consuming Places*, London: Routledge.

Urry, J. (2002) *The Tourist Gaze*, 2nd edn, London: Sage.

van den Berghe, D. and Keys, C.F. (1984) 'Introduction: tourism and re-created ethnicity', *Annals of Tourism Research*, 11(3): 343–52.

Wonders, N.A. and Michalowski, R. (2001) 'Bodies, borders and sex tourism in a globalized world: a tale of two cities – Amsterdam and Havana', *Social Problems*, 48(4): 545–71.

Wood, R.E. (1998) 'Touristic ethnicity: a brief itinerary', *Ethnic and Racial Studies*, 21(2): 218–41.

Wood, R. and Picard, M. (1997) *Tourism, ethnicity and the state in Asian and Pacific societies*, Honolulu: University of Hawaii Press.

8 Tourism and New York's ethnic diversity

An underutilized resource?

Susan S. Fainstein and John C. Powers

The observation that 'tourism is caught in a paradox whereby its success can destroy it' has become a commonplace. As the industry grows ever larger, it seeks to make places simultaneously interesting to visitors as well as safe and comfortable. Competition among cities to become destinations for travellers forces them to develop facilities that are increasingly indistinguishable from one another, causing the weary traveller to wake up and wonder if she is in Brussels or New York. Luxury hotels, convention centres, sports venues, concert halls, museums and shopping malls have a sameness not easily overcome by architectural embellishment, winning teams or fancy shop windows.

Many cities are addressing the dilemma of how to convey uniqueness by developing and marketing parts of the urban landscape that were previously considered marginal or dangerous. In Europe these are immigrant quarters; in the United States they are both immigrant enclaves and African-American ghettos. This marketing of diversity fits well with the post-Fordist notion of niche marketing and customized production. At the same time, however, US cities are inhibited from investing too heavily in this strategy by the financial power of the establishments that constitute the main tourism interests. Moreover, the very diversity of the neighbourhoods that require marketing limits the extent to which they can effectively be advertised. Since each is different, each calls for a separate effort but lacks sufficient resources to promote itself.

Discussions over a tourism strategy that emphasizes diversity veer between two views. The first criticizes governing regimes for directing their energies toward central business districts and large corporations at the expense of community development and ethnic entrepreneurs. The second reflects fears that promoting tourism to ethnic neighbourhoods will result in commodification and commercial gentrification. We argue in this chapter that New York's efforts toward directing tourists to outlying neighbourhoods are negligible, and that the dangers of socially excluding minority groups exceed the dangers of commodification. The continued focus of New York's capital expenditures on convention centres and sports venues, coupled with the weakness of the diversity marketing effort put forth by the city's principal tourism agency, perpetuates the Manhattan-centric policies that have historically guided New York's urban redevelopment programmes (Fainstein and Fainstein 1989). Thus, we view the investment approaches and tourism services proffered by New York's tourism regime as largely contributing to the ongoing exclusion of low-income neighbourhoods from the economic and cultural life of the city.

Social exclusion

The term *social exclusion* has become widely used in European discourse, less so in US policy discussions:

> These processes [of new patterns of unemployment, poverty and deprivation within Europe] have been accompanied by a shift in both policy and academic discourse from a dominant conception of poverty to a focus on social exclusion, signifying a significant redirection of emphasis from the material deprivation of the poor towards their inability to fully exercise their social, economic and political rights as citizens. ... Social exclusion signifies the combined impact of factors such as lack of adequate education, deteriorating health conditions, homelessness, loss of family support, non-participation in the regular life of society and lack of job opportunities. Each type of deprivation has an impact on the others. The result is seen as a vicious circle.
>
> (Geddes 2000: 782–3)

To a large extent there is a parallel between this European usage and the application of the term 'social isolation', as used by William Julius Wilson (1980, 1990) in his analyses of the US ghetto. In both European and North American discussions, the proposed remedy is a better integration of the socially excluded into the urban fabric.

Social exclusion, as a concept, hence refers to spatial segregation but also to detachment from social institutions or consignment to inferior ones. While physical isolation helps to reinforce social exclusion, place-based remedies alone are inadequate to address processes of class and racial/ethnic stratification. Yet place-based programmes can interact with other benefit programmes to forge conditions for a higher quality of life for low-income people, and can take into account more than purely material conditions (Markusen and Fainstein 1993). Improving opportunities in isolated urban neighbourhoods, and destroying the perception of them as 'no-go' areas, are needed complements to redistributional programmes, and can help to break down prejudices that inhibit other forms of assistance. It is here that increased tourism can play a role by changing urban imagery, causing places formerly regarded as scary to seem interesting or edgy and daring instead.

It has long been stylish to deride tourists as destroyers of culture (see especially MacCannell 1976) and to regard festivals produced for visitors as inauthentic spectacles (Debord 1994; Harvey 1990; see Judd 2003 for a counter-critique). But it is problematic to condemn visitors for displaying a repugnant orientalism just for regarding the quotidian lives of disparate people as exotic. Doubtless all social groups weave fantasies about 'the Other' (as well as myths about themselves); whether or not such fantasies are harmful depends on situations and outcomes that cannot be prejudged. In contrast to the prevailing cynical tone, there has also been a hopeful thread in tourism debates, identifying travel as a source of educational benefits and a means of promoting understanding among culturally different groups. Which tendency prevails – malignant stereotyping or growing tolerance – is largely context-dependent, and even then both processes can operate simultaneously (Fainstein and Gladstone 1999). Also context-dependent is whether commodification that leads to commercial and residential displacement by large corporate interests actually presents a serious threat to ethnic enclaves, particularly in a global city like New York. As we discuss below, this has not been a major problem so far, except in Harlem and in central Brooklyn. Thus, whether diversity tourism demeans or whether its absence excludes, whether it improves economic opportunity or displaces local businesses, can only be determined empirically and situationally.

Tourism, like other economic sectors, is governed by a regulatory framework that structures the context in which business transactions occur. This framework encompasses the formal rules imposed by institutions as well as the informal norms and expectations that arise from social and cultural patterns (Amin 1994). Usually applied to the national level, arguments about regulation can be extended to the urban milieu (Hoffman *et al.* 2003; Lauria 1997). The impact of tourism on urban neighbourhoods depends to a great extent on the regulatory system in which the tourism operates. Case examples in Harlem and Brooklyn discussed below indicate how the distribution of the economic benefits of tourism development can be affected when policy choices within a structured system of regulation are contested and influenced by mobilized local groups.

This chapter first sketches the geographical context of tourism in New York City, and then describes and evaluates the city's limited programmes for tourism promotion outside the Manhattan core. It depicts some of the organizations and schemes that have tried to attract visitors based on the unique characteristics of ethnic neighbourhoods, identifies obstacles to their success, and concludes by discussing the relationship between tourism, diversity and redevelopment.

New York's geography

Efforts to direct visitors to New York's ethnic neighbourhoods run up against a stubborn and intense focus on Manhattan, which itself results from the physical and perceptual geography of the city. Encompassing over

400 neighbourhoods and roughly 180 different languages, New York City exhibits a unique variety of community and cultural assets, all of which could be available to the determined visitor. Consisting of five boroughs – Manhattan, Bronx, Brooklyn, Queens and Staten Island – the city is home to just over 8 million people, according to 2000 census figures. Brooklyn and Queens are the two most populous boroughs, respectively numbering about 2.5 million (31 per cent of the population) and 2.2 million (28 per cent). Manhattan is home to some 1.5 million (19 per cent) of the city's residents, followed closely by the Bronx with 1.3 million (17 per cent). Each borough is parsed into subunits called community districts, each containing several neighbourhoods (see Figure 8.1). New York City has 59 community districts, only 12 of which are in Manhattan. The areas posting the largest population gains over the past 10 years were mainly in the 'outer boroughs' (i.e. outside Manhattan), namely in the central Bronx,

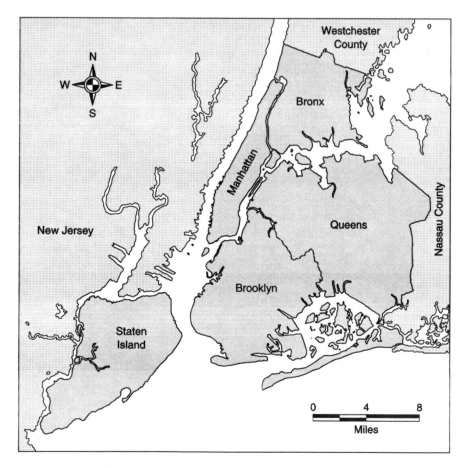

Figure 8.1 Neighbourhoods in New York City, 1990

south-eastern and south-western Brooklyn, and virtually all of Queens and Staten Island (New York Department of City Planning 2004).

The overwhelming share of the city's non-white and immigrant population resides outside of what is often called 'Manhattan south of 96th Street'. Owing in large measure to the 1.2 million immigrants that arrived in the city over the 1990–2000 period, communities in northern Manhattan and the outer boroughs have changed dramatically. For instance, various parts of Queens such as Elmhurst, Flushing, Jackson Heights/North Corona, Sunnyside and Woodhaven/Richmond Hill have all experienced a sizeable influx of Hispanic and Asian residents, altering the ethnic and racial characteristics of the neighbourhoods (Figures 8.2 and 8.3). In Brooklyn (home to some of the city's highest-quality housing stock), neighbourhoods such as Fort Greene, Canarsie and Sunset Park/Windsor Terrace have experienced influxes of both non-white immigrants and

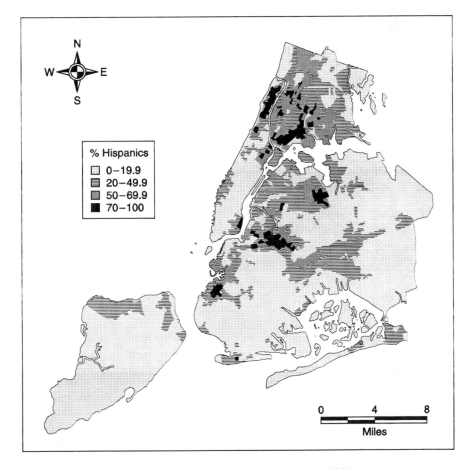

Figure 8.2 Hispanic neighbourhoods in New York City, 2000

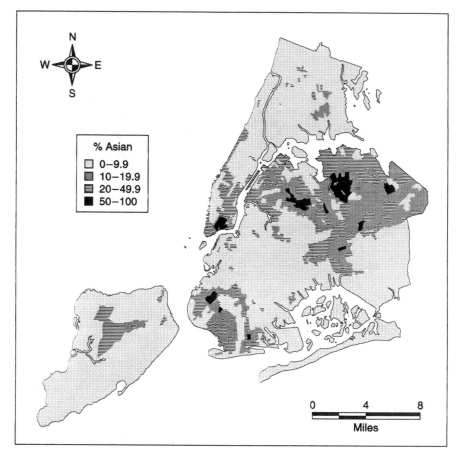

Figure 8.3 Asian neighbourhoods in New York City, 2000

black and white gentrifiers. New construction and rehabilitation is occurring in some of these areas, as well as in other neighbourhoods such as
Morrisania/Corona Park East and Hunts Point/Longwood in the Bronx
and East Harlem in Manhattan, where many second-generation or more
established ethnic Hispanics have begun to settle (New York Department
of City Planning 2004).

Building stock in these areas varies widely, with mixes of town houses,
rows of tenements, houses large and small, privately owned apartment
buildings and public housing projects. Generally the street pattern consists
of broad avenues lined with storefronts, with narrower, residential streets
branching off. A wide array of ethnic small businesses is present.
Restaurants, music and dance clubs, clothing stores and jewellery boutiques,
as well as travel agencies, money transfer institutions (for sending remittances) and, in some cases, small hotels, are examples of businesses that

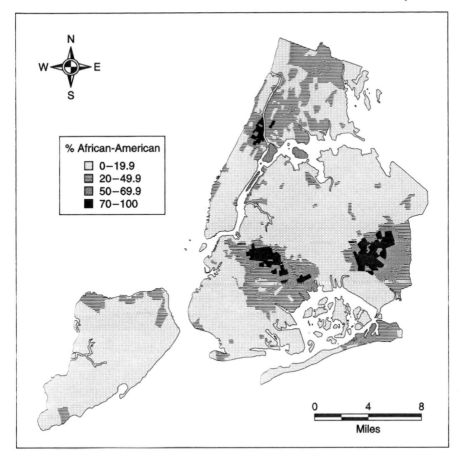

Figure 8.4 African-American neighbourhoods in New York City, 2000

serve local clienteles while also forming important linkages to business suppliers and distributors outside their immediate locales. They are patronized by visitors from both inside and outside the metropolitan area. In some cases, new business expansion has generated a variety of new services and amenities for both recent and long-standing residents. The view from within many of these neighbourhoods is one of a diverse, and sometimes very dynamic, local base.

Access to these areas, however, is daunting for the average tourist, who stays in a midtown Manhattan hotel, feels intimidated by the complex subway system, and can expect little information about neighbourhoods outside midtown and downtown Manhattan from hotel concierges or other standard sources.[1] Although the outer boroughs possess cultural resources that would be veritable magnets for visitors in other metropolises, it is extremely difficult for them to compete with Manhattan attractions.

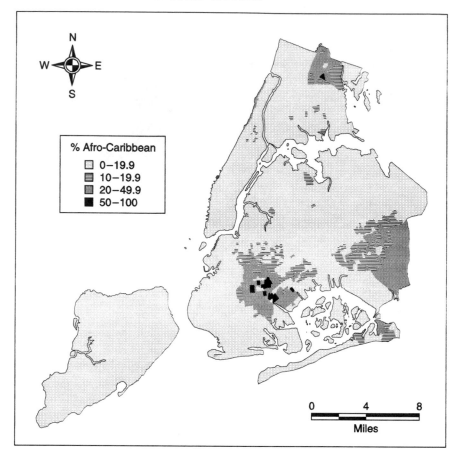

Figure 8.5 Afro-Caribbean neighbourhoods in New York City, 2000

Tour buses do serve Harlem (in Manhattan, but north of 96th Street) and recently Brooklyn, but most ethnic and minority communities have not even been put on the maps distributed to tourists at hotels and visitors' centres. While all the boroughs have very large populations (only Staten Island has under a million people), the pull of central Manhattan overwhelms them. Most of the major museums, all the principal hotels, the highest-rated restaurants and the most famous landmarks are all in Manhattan.

Furthermore, the economics of hotel investment linked to property development put many neighbourhoods in Harlem and the outer boroughs at a structural disadvantage for attracting tourists. Because hotel owners have no vested interest in encouraging tourism outside of their own vicinity, they provide no impetus for tourism development in ethnic neighbourhoods.[2]

The wide diversity of ethnic restaurants scattered about the city would benefit greatly from more visitors, but most of them are small-scale, 'mom-and-pop' operations, whose proprietors have little spare time or cash to invest even in joint marketing ventures. They must depend on the paltry municipal efforts to promote tourism outside the Manhattan core.

Individual travellers inclined to strike out on their own are inhibited, too, by the chronic dearth of information in guidebooks to the city, which are also extremely Manhattan-centric. In the most popular ones (*Eyewitness Travel Guide*, *Fodor's New York City*, *Time Out New York* and *Frommer's New York City*), only three ethnic or minority neighbourhoods in Manhattan (Chinatown, Little Italy and Harlem) rate serious coverage. While *Time Out* and *Fodor's* do pay some attention to ethnic restaurants in Brooklyn and Queens, *Frommer's* and the *Eyewitness Guide* focus almost solely on Manhattan south of 96th Street. Even the *Lonely Planet Guide*, aimed at the more adventurous traveller, restricts itself largely to the same territory, though it does recommend riding the No. 7 subway train ('the international express') to Queens. For those seeking more information, there are books devoted just to Brooklyn, as well as a book entitled *New York Neighborhoods*, aimed entirely at those wishing to sample ethnic cuisines and shops. These are not widely distributed, however, and they tend to get out of date.

Marketing New York City – NYC & Company

City marketing in the United States is typically carried out by hotel and convention bureaus, funded by subventions from the hotel and restaurant industries. The major players in these industries expect the bureaus to devote the bulk of their resources directly to their interests. In New York City – North America's, and maybe the world's, most diverse metropolis – the city's convention and visitors bureau named NYC & Company receives just $6.5 million (€4.5 million) in public funding, far less than the city collects from its hotel tax. Its primary funders are its subscribers ($8 million), who are mostly hotel and restaurant owners.[3] As a consequence, its promotional activities centre on attracting large conventions, staffing a midtown Manhattan tourism office, wooing travel agents, and advertising events and tourism packages. Virtually all the leaflets available in its travel centre focus on sites and happenings in Manhattan south of 96th Street, and so does its website. Historically, the urban regime has seldom viewed districts outside the Manhattan core as assets, and much of the attention they did ever receive was in the form of social programmes. Few people envisaged them as tourist attractors.

NYC & Company is a public–private partnership with an independent board of directors that appoints its chief executive.[4] Even though it is the official tourism promotion body of the city, it does not report to the city government. It does work closely with the city's Economic Development Corporation. Its 2003 operating budget of $14.5 million was meagre next to those of comparable bodies in other major tourist destination cities in the United States (such as the Las Vegas visitors bureau with its $140 million annual budget for tourism promotion). NYC & Company has two basic parts to its operations. The first focuses on attracting large conventions and exhibitions and involves the wooing of meetings organizers; the second supports major events and promotions like the US Tennis Open and the Macy's Thanksgiving Parade. Although the organization perceives itself as oriented toward the city as a whole, its efforts are systematically targeted at high-rate-of-return types of activities concentrated in Manhattan. Its corporate backers expect it to demonstrate a return on any advertising investments it makes in large-circulation magazines or television. This compels a marketing approach highlighting the tourism venues, corporate gatherings and major events that best exploit Manhattan's comparative cultural advantage for attracting visitors.

As the city's official marketing arm, NYC & Company does try to act as a 'gateway' to promotional activities undertaken by organizations in other parts of the city. Recently, as we shall see, numerous attempts have been made to develop tourism in Harlem and outer boroughs, especially Brooklyn and Queens. Efforts have been mounted by borough presidents' offices[5] and by non-profit organizations concerned with economic development in poor neighbourhoods. These are part of a general move to revitalize commercial areas outside the core business district. It is still too early to judge their success, except impressionistically. Ethnic groups generally seen as socially excluded are territorially peripheralized. As in most European cities, the elite of New York occupy the city centre, pushing immigrants and African-Americans into the 'outer boroughs' and northern Manhattan. This results in part from a steep gradient in land values as one moves out from the centre, but it is also a consequence of deliberate policy interventions, in the form of urban renewal, highway and neighbourhood revitalization programmes aimed at removing low-income households from central locations (Fainstein and Fainstein 1989). Since Manhattan is an island surrounded by wide rivers, activities pursued there do not bleed easily into the surrounding territory. Although parts of Queens are just a brief subway trip from midtown Manhattan, and downtown Brooklyn is similarly close to downtown Manhattan, the East River still forms a great psychological barrier. Furthermore, many of the interesting ethnic areas in these boroughs, like Flushing in Queens and Brighton Beach in Brooklyn, require much longer journeys. For those opting to use their car or take a taxi, traversing the river can be a major undertaking due to traffic back-ups at bridges and tunnels. Getting to the Bronx from midtown is a quite long subway ride, and Staten Island involves either a half-hour ferry ride or a circuitous route through Brooklyn and across the Verrazano Narrows Bridge.

Umbrella marketing and the targeting of neighbourhoods

Since the terrorist attacks of 11 September 2001, NYC & Company has shifted some of its attention from the heart of the tourism district in midtown to neighbourhoods in lower Manhattan that suffered a loss of business after the attacks. Making use of federal allocations for the revival of the downtown area, it launched an advertising campaign for Chinatown. Mostly, though, it has sought to increase community participation in the marketing of neighbourhood events, such as by organizing local businesses as sponsors, helping with content and programming, and providing 'umbrella marketing' of surrounding neighbourhoods during cultural activities and festivals. Examples of two such campaigns are 'Christmas in Little Italy' and the 'Chinese Lunar New Year'. NYC & Company has used these time-bound interventions to highlight the cultural uniqueness of the two neighbourhoods (both in lower Manhattan) and to encourage spillover effects into adjacent communities. It remains to be seen how much success such campaigns can achieve if restricted to infrequent cultural events. Perhaps the strategy could be expanded to market the cultural diversity of specific neighbourhoods more comprehensively.

The assumption that everyone knows about New York anyway, and that the city gets enormous free publicity from the mass media, has been used by city administrations to justify budgetary stinginess in matters of tourism. Because NYC & Company is constrained by its tiny operating budget, it seeks to capitalize on this 'brand recognition'. Yet most portrayals of New York in film and television, with the exception of crime stories, focus once again on the glamour of Manhattan. NYC & Company has made only very minor efforts to work with various business improvement districts (BIDs) or community organizations to expand tourism beyond the core portion of Manhattan. It distributes small seed grants to specific neighbourhoods in the outer boroughs and it designs sample itineraries for city tour operators in unique areas of the city,[6] but it allocates only $30,000 annually to each borough for tourism promotion.[7] Even when major events are held in boroughs, the agency does not work to ensure neighbourhood spin-offs. Thus, in conjunction with the US Tennis Open in Flushing, Queens, NYC & Company once promoted an associated restaurant discount scheme in which *all* the dining places were in Manhattan, despite Flushing's plethora of Asian eateries. A *New York Times* article[8] focused on the plight of that neighbourhood's commercial establishments:

> John Guo, who works at Ten Ren Tea House, said he would like to see more people venture from the sports sites to his neighborhood, but few make the trip [from either the tennis venue or from nearby Shea Stadium, home of the New York Mets baseball team] 'Nobody seems to know we are here,' he said.

NYC & Company perceives tourism expansion within the outer boroughs as the responsibility of the borough presidents, each of whom has a staff member responsible for tourism development. Although this approach avoids any duplication of effort, it effectively limits the resources.

Other organizations

The high concentrations of immigrant and ethnic groups in New York, the variety of their cultural and religious characteristics, and the dense and mixed nature of communities living in close proximity to one another has propelled some organizations and institutions to market their communities and to publicize the cultural assets inherent in these diverse surroundings. Typically, their principal focus is not just tourism. They act more as neighbourhood advocates, working to preserve local culture and define a niche for their areas within the larger metropolitan economy. Their work is presented briefly here.

Offices of the borough presidents

The common perception within most borough presidents' offices is that city government does precious little to promote the diverse, ethnically mixed areas outside Manhattan below 96th Street. In addition to sponsoring projects in their own boroughs, these offices play the lead role in marketing the communities and businesses located there. They have recently launched a joint effort called More NYC. Aware that the cultural and ethnic richness of the boroughs is one of their greatest assets, borough presidents increasingly try to seize on these attributes to realize economic advantages. However, the resources they have available to give shape and meaning to their plans do not necessarily match their commitment. Many of the staff responsible for tourism have to divert time from their other, primary tasks.

Because the motives and backgrounds of borough visitors differ from those of the typical tourist, different forms of marketing are required. Much tourism comes to the boroughs in the form of extended family reunions, weddings and cultural festivals. Family members may arrive from South Asia (India, Pakistan, Bangladesh, Burma), from Central America and the Caribbean (Dominican Republic, Puerto Rico, Mexico, El Salvador), from Eastern Europe and Russia, and from other corners of the globe. These visitors often come for extended stays (several weeks). Although they generate little hotel business because they reside with family members, they do create considerable demand for local shops and restaurants. In areas like the South Bronx and Jamaica, Queens, the origins of the hip-hop culture also attract substantial inbound tourism from places as far away as Germany and Japan. This has given rise to numerous dynamic market outlets for clothing and music stores.

International and out-of-town travellers to ethnic neighbourhoods often intermingle there with native New Yorkers and suburbanites who come back to 'the old neighbourhood' for shopping and dining.[9] Queens and Brooklyn offer a rich variety of cultural assets, including museums (the Brooklyn Museum of Art, home of a major collection of European and Asian fine art, and the Museum of African Art, among many others);[10] major television studios in Astoria, Queens; and some of America's first planned communities, such as Forest Hills Gardens and Sunnyside Gardens in Queens. Visitors to these sites may also be attracted to the large concentrations of ethnic specialty stores and restaurants nearby. In Brooklyn, neighbourhood gentrification in places like Prospect Park, Fort Greene, Greenpoint and Williamsburg is remaking the image of these old residential areas. This attracts new investment in nightclubs and restaurants and brings in sizeable numbers of new visitors, particularly from Manhattan. Though the trendy clubgoing set has many here-today-gone-tomorrow attributes, its recent arrival in Brooklyn is nevertheless a sign of neighbourhood change linked to a new market dynamics one created by customers from other parts of the city and with a potential to attract visitors even from outside the metropolitan area.

The borough presidents' offices have acknowledged the need for coordinated efforts among a wide range of actors. Many different initiatives are being attempted, especially along the lines of informal tourism coalitions that bring together community and cultural associations, local hotels, travel agencies, local development groups and BIDs. The Queens office is working toward formal coordination among different stakeholders and has held a borough-wide conference on tourism development. The Bronx Tourism Council has set more targeted objectives, particularly involving hotel development, which is severely lacking in the borough. In Brooklyn, the emphasis is on getting people to explore different neighbourhoods. The borough president has just opened a 1300 square foot (120 m^2) tourism centre in Borough Hall, the first such facility in any of the outer boroughs.[11] This endeavour is highly significant, given the current fragmentation of the information provision about sites to visit, transport options and other key issues for visitors like expense, safety and accommodations. While the centre in itself will not entice visitors from Manhattan to Brooklyn, it will at least guide them once they get there, as well as assisting outsiders visiting friends and relatives. One problem is how to make potential clients aware of the facility, which is not located in a part of Brooklyn where visitors are likely to arrive.

It is important to stress that the boroughs discussed here – Brooklyn, Queens and the Bronx – encounter different obstacles to the expansion of tourism and the marketing of their unique community and cultural assets. For instance, Queens, home to the two New York airports, has the largest hotel stock outside of Manhattan, while the Bronx has the smallest.

Nonetheless, all borough presidents face certain common obstacles, too, which impede their efforts to expand tourism. The lack of signage is often cited as a major factor restricting both informational and logistical access to places of interest. Borough staff members have stressed how the poor signposting contributes to a lack of a sense of place and creates psychological barriers to exploring particular neighbourhoods, especially as a pedestrian. This is a key factor limiting the commercial viability of the hop-on hop-off bus tours that have been extended to Brooklyn and Harlem (but that still do not visit the remainder of the city outside central Manhattan).

Inadequate transportation is a complex issue, but one that is universally cited as a fundamental constraint on the accessibility of various neighbourhoods, and indeed of whole boroughs. Although the airports are in Queens, most connections from them go to Manhattan, and both railroad stations are in Manhattan. Since the subway system was mainly designed to carry workers from outlying areas to Manhattan, there is little interconnectedness within the boroughs themselves. Water-based transit is also poorly developed. The popular Circle Line cruises carry passengers around Manhattan, and new ferry links bring commuters to the two main Manhattan business districts. Private car transport is hampered by the scarcity of parking space, yellow cabs do not cruise the streets of outer boroughs[12] and visitors to New York are less likely to rent cars than their counterparts in other, less centralized US cities.

Consciousness-raising within neighbourhoods with regard to tourism remains an issue at multiple levels in the outer boroughs. Borough staff concede that many local businesses, not to mention certain branches of local government, do not display the types of attitudes and basic behaviours that tourists find welcoming. To improve the marketing of the boroughs, one also needs to instil a general understanding of what it *means* for a community to be open to visitors. The attempt by the Queens borough president to convene major stakeholders to discuss tourism strategy can perhaps be best understood in this light.

Borough-level governments increasingly take the lead in devising strategies of economic development that assign key roles to tourism. But because the borough offices depend on the central city budget for their existence, they are starved for the resources they need to support tourism development. Moreover, many of the problems they face can be resolved only at higher levels of the planning, transportation and economic development apparatus of New York City government. Nevertheless, the manner in which boroughs are forging new public–private relationships with community and cultural associations is testimony to their creative attempts to capitalize on the wealth of cultural and historical assets they have to offer. Other local actors are giving voice and meaning to what these assets are, how to preserve them and how to bring them to the attention of a wider range of interested people.

City Lore and the Place Matters project

City Lore is a non-profit organization initiated in 1986 by a folklorist, originally from the Smithsonian Institution in Washington, who was dedicated to preserving the living cultural heritage of New York City. In recent years, City Lore has teamed up with the Municipal Arts Society of New York to launch a project called Place Matters. The project is more than just an attempt to save historical and cultural landmarks from the claws of property development. It seeks to establish an ongoing recognition of how current natural and built city environments are linked to rich layers of history, community and memory. Such knowledge is considered an enhancement to urban living. City Lore and Place Matters work proactively to protect many physical landmarks, as well as the cultural practices that have grown up around such landmarks in past and present. In this way, their advocacy work distinguishes itself from many other preservation campaigns, which often merely react to specific crises and focus solely on buildings rather than on social history and culture.[13]

Expanding tourism is therefore not City Lore's primary objective, though the organization does perceive tourism development as helping people to think about place in new ways. It sees itself as an advocate for the culture that pervades physical landmarks. Place Matters has conducted a Census of Places That Matter – sites in New York that embody the richness and diversity of its history. Although the project believes that tourism can add value to the cultural assets of a particular site, the knowledge bank resulting from the census is primarily meant to heighten awareness of the importance of locale. Place Matters has teamed up with various community development corporations (CDCs), particularly in the Bronx, to stimulate neighbourhood tourism as a means of empowering communities and preserving expressive cultures. Such tourism efforts, however, are as much about the *unique character of place* as about the architecture associated with unique landmarks.

Neighbourhood tourism and walking tour organizations

Quite apart from the sights at which most New Yorkers and city visitors marvel on a daily basis, such as the city's architecture, museums, monuments and showplaces, New York's extraordinary diversity is perhaps best understood in interactions at a neighbourhood level. Several organizations are actively introducing New Yorkers and visitors to a broad array of neighbourhoods whose historical and cultural roots are still shaping neighbourhood change. These groups are particularly sensitive to issues of authenticity, and they try to provide an accurate picture of the neighbourhood. Precisely for this reason, however, they may do little to stimulate economic activity and may themselves be economically shaky.

Walking tour organizations include community and historical associations in the cases of Fort Greene in Brooklyn and Richmond Town in Staten Island; in various parts of Queens and Brooklyn they include BIDs; and in the South Bronx they include CDCs. One recent significant effort was funded by the Conference Board, an elite business group, but it foundered when its executive director took another job. Even before that, it had serious difficulties attracting participants. It lacked visibility because of an inadequate advertising budget. The types of difficulties faced by such organizations are exemplified by one such walking tour, which chronicled the musical history of the South Bronx from the mambo era of the 1950s through the rise of hip-hop in the 1970s. An arts-based CDC called The Point sponsored the tour, in which the cultural origins of the Dominican, Puerto Rican, Cuban and African-American populations were interwoven in an understanding of the neighbourhood's musical history. The CDC could not allocate dedicated staff to the project, but relied instead on the Conference Board to pay a tour leader on an ad hoc basis. When the board failed, the CDC's backing also evaporated.

In Fort Greene, a predominately African-American and African-Caribbean neighbourhood in Brooklyn, businesses have not focused on tourists per se, but have tried to attract outsiders drawn to the neighbourhood by walking tours and by events at the Brooklyn Academy of Music. The walking tours are sponsored by the Historic Fort Greene Association and they bring in both New Yorkers and visitors from farther afield. Though the tours focus on architecture and ethnography, the organization also distributes lists of all businesses in the area to people on the tours. In response, businesses are trying to capitalize on the presence of outsiders; for example, restaurants are opening for brunch on tour days.[14]

Many of the visitors to this black neighbourhood, like those travelling to immigrant neighbourhoods, are of the same ethnicity as the residents. According to one observer,

> In a locally owned shop that sells body care goods, I overheard customers appreciate the establishment because they perceive that the products were designed for a predominately black customer base.... When I spoke with Michelle, the shop owner, she stated: 'Some of my customers come all the way from Atlanta. They come to shop and hang out in the area.... They tell me they can't get "this" in the South' – referring to the products but also the neighbourhood's cultural capital associated with the fusion of people, foods and products inspired by black peoples from the Caribbean, West Africa and America. The 'authenticity' of the neighbourhood hinges on its black Diaspora representation, its class differences and its integration of artists and business enterprises.
>
> (Sutton 2002)

These case examples lead us to conclude that the marketing of diverse neighbourhoods does not necessarily constitute a commodification of Otherness or a presentation of exoticism for the benefit of the white bourgeois consumer. Yet precisely this fact makes it difficult to locate and advertise to the potential customer base.

The core set of obstacles to a more viable tourism economy includes both contingent and more structural factors. Marketing efforts are intermittent and are not linked to larger tourism promotion structures, such as those offered by NYC & Company. Tour guides carry out tours on an informal basis, as demand is unreliable. Demand could be stabilized by appealing to larger numbers of foreign visitors arriving every year to seek the cultural roots of popular music (at present especially those of the hip-hop scene). But these visitors are constrained by the lack of accommodation options outside central Manhattan.

The potential for diversity tourism

The efforts documented above are sporadic and highly dependent on individual initiatives, rather than being based in an established institutional structure. New York City's peculiar geography, the concentration of entertainment and retail activity in a strikingly small area, the traditional bias of city government toward Manhattan, and its long-time, rather anomalous underfunding of tourism promotion all inhibit the development of tourism in minority and ethnic neighbourhoods. Unlike many cities (such as Miami, Toronto or London) where ethnic neighbourhoods border directly on the central business district, those in New York are further flung and not easily accessible, with the exception of Manhattan's Chinatown and Little Italy. In many European cities, by comparison, where principal sights like churches and museums are similarly scattered about the city, public transportation is more readily available and hotels are much less concentrated in city centres.

Despite all this, the potential does exist in New York to increase the numbers of visitors to lower-income neighbourhoods. Harlem, once considered a no-go area, has recently been attracting 800,000 tourists a year. Whereas not long ago it offered few sights beyond the Sunday morning gospel services at its many churches, it has since spawned a number of restaurants and jazz clubs to meet the new demand. Another attraction there is the revived Apollo Theater, a historic incubator of talented African-American performers (Hoffman 2003). Brooklyn, home to nearly 2.5 million people, did not have a single first-class hotel until recently. When a developer tried to build one, banks initially refused financing, considering Brooklyn too risky. Once he finally succeeded in building the hotel, now operated by the Marriott chain, he achieved almost 100 per cent occupancy and is currently launching an expansion. Construction of another new hotel in Brooklyn is now also underway. Growing membership of

Harlem businesses in NYC & Company has drawn more attention to that area, too, and a hotel is also under construction. These successes illustrate the possibilities that exist if entrepreneurs come along to exploit them.

Yet progress is slow. Many of the obstacles summed up here still impede efforts at business development. In sharp contrast to the weak governmental support for efforts to improve the prospects of local small businesses, both the city and state governments direct massive resources at mega-developments to get out-of-towners to visit New York. And even when the city proposes projects for outlying boroughs, they do not represent real changes in the direction of public policy. Furthermore, it is doubtful in many cases whether adjacent immigrant or African-American communities would actually stand to benefit.

An example is the Brooklyn Atlantic Yards project, a vast scheme whose centrepiece is to be a new basketball arena for the New Jersey Nets franchise. The project would transfer this professional team from its current home in northern New Jersey to Brooklyn, in the wake of its purchase by Brooklyn-based property developer Bruce Ratner. The project, to be built by his Forest City Ratner Companies above existing railroad yards, also extends into neighbouring areas, causing both residential and commercial displacement. From the developer's point of view, the basketball arena will be ancillary to the more profitable parts of the project, which include a 21-acre (8.5 ha) mixed-income residential and commercial development, some 1.9 million square feet (176,500 m^2) of class A office space, and associated commercial development. City government sees the project as the resurrection of downtown Brooklyn.

Our intention in discussing this scheme is not to undertake a cost-benefit analysis, but to show how this public–private marshalling of resources and investment capital exemplifies the city's current economic development efforts and their likely effects on the diversity of place-bound economic and cultural assets. In other words, we are concerned that this type of investment strategy, even when directed at areas outside Manhattan, merely pre-empts other approaches to the economic and sociocultural enhancement of the more diverse urban communities and neighbourhoods in a global city like New York. To assert that the city will greatly benefit from new revenue sources generated by mega-developments is not the same as saying that Brooklyn residents and business owners will reap either direct or derived benefits from increased commercial activity and foot traffic in the area. The generous subsidies and wide-reaching political influence available to major developers, and in particular the excessive government support given to sports teams, contrast starkly with the neglect of small-scale efforts by local ethnic entrepreneurs. Any discussion of the Atlantic Yards project's relative merits should take full account of the particular *brand* of cultural consumption and production being pursued, and of the trade-offs required by the project within the urban economy. Ultimately this project, if consummated, will bring in chain stores and raise rent levels in its environs. While it may indeed provide some housing and employment for Brooklynites, that will be at the expense of diversity and will hamper the promotion of Brooklyn's unique attractions. The enthusiasm for vast investments in sports facilities contrasts with the feeble efforts to invigorate ethnic businesses. Their economic status is fragile, and their potential for further development is forever being threatened by the appeal that mega-projects enjoy within New York city government.

Be careful what you wish for

The distinctiveness of many ethnic communities means they offer the tourist something qualitatively different from the standard Broadway- and Manhattan-based experience. It is yet unclear how big the market could be for large-scale tourism in neighbourhoods like Jamaica and Jackson Heights in Queens, Fort Greene in Brooklyn, the South Bronx, and many others, and what threats such tourism might pose to the authenticity of specific neighbourhood experiences. The problem of defining the market for this place-based tourism – and of predicting what it might mean in terms of loss of authenticity, displacement of small-scale businesses, and residential gentrification – raises the issue of what the gains from tourism actually are and how they can be translated into community well-being. Must place-based tourism exist at a low-level equilibrium to be sustainable for the community that plays host to the visitors? Or does tourism promise improved amenities for any marginal neighbourhood and an expanded customer base for its local businesses?

Tourism is a double-edged sword, and the likely answer to the questions above is that both outcomes are possible. Some of New York's ethnic enclaves have already turned into artificial efflorescences, embodying a nostalgic image with little connection to contemporary reality. Thus, Little Italy in southern Manhattan and the Bronx's Arthur Avenue (another Italian restaurant and shopping enclave) both continue purveying images of their Italian origins long after the Italian residents have departed. Little Italy restaurants feature promoters who stand in front of the restaurants exhorting passers-by to sample their indifferent cuisine. Arthur Avenue, which has maintained higher culinary standards, caters mainly to suburbanites retracing their New York roots; it is surrounded by a Hispanic residential population. Manhattan's Chinatown, in contrast, despite its long history of serving outsiders in its countless restaurants, is still the first stop for new immigrants and still boasts many shops and eateries catering to Chinese-speaking residents and visitors. Its average income remains very low, and its diverse economy includes garment factories, a variety of retail, wholesale and service establishments, banking institutions and a flourishing informal sector. Harlem, though threatened by both residential and commercial gentrification, has strengthened its commitment to black culture and, after witnessing a decline of its music and food scene, has generated a variety of black-owned jazz and eating places (Hoffman 2003).

Tourism thus has the potential to liven up many New York neighbourhoods and to strengthen their appeal to residents and visitors alike. Tourism is synergistic with the creation of amenities that also benefit residents. It can boost the demand for local culture, refurbished architecture, attractive parks and distinctive businesses. Achieving development based on diversity rather than on the exploitation of difference, stimulating cultural interchange while avoiding putting local people on exhibit, nurturing small business while keeping chain stores at bay – all these are possible benefits of opening up neighbourhoods to tourists. The trick is to secure the benefits while minimizing the obvious pitfalls.

Notes

1 It is alleged that hotel concierges are rewarded financially for directing guests to expensive Manhattan restaurants. Small proprietors in the boroughs cannot match this kind of initiative.
2 A new hotel now being built in Harlem, and recent plans for another, may slightly alter this situation.
3 Information supplied by NYC & Company.
4 The discussion of NYC & Company is based on interviews with NYC & Company staff and with knowledgeable informants.
5 The elected presidents of New York's five boroughs have limited executive powers over services and capital expenditures in their jurisdictions. Most activities are funded through the mayor's office, but the borough presidents do have a limited discretionary budget. There are no borough legislative bodies.
6 The chief executive of NYC & Company, in response to an article implying criticism of the organization's meagre efforts at tourism promotion in the boroughs, responded that 'promoting and increasing tourism for each borough is a cornerstone of NYC & Company's mission'. She cited its sponsorship of Culture Fest, a meeting to market arts and cultural organizations from all five boroughs. (C.L. Nicholas, letter to the editor, *Crain's New York Business*, 13 September 2004). Significantly, the meeting itself took place in the downtown Manhattan business district.
7 'Boroughs act for attention', *Crain's New York Business*, 6 September 2004.
8 L. Polgreen, 'Ethnic food, anyone? After tennis matches and baseball games, Flushing beckons', *New York Times*, 2 September 2004.
9 For instance, at 74th Street in Jackson Heights in Queens, Indian merchants sell a highly sought-after type of gold that is used in Indian weddings and is difficult to source elsewhere in the metropolitan area (personal communication, Director of Economic Development, Office of the Queens Borough President, 10 October 2003). Many ethnic cooking ingredients unavailable outside New York can be bought in such neighbourhoods too.
10 During the remodelling of the Museum of Modern Art's landmark building on 53rd Street in Manhattan, it displayed much of its regular collection and mounted several major exhibits in a converted factory in Long Island City, Queens, just across the river from Manhattan. Traffic to the museum provided business to local restaurants and galleries. When the museum returned to its

permanent Manhattan home in November 2004, however, it closed its Queens facility to the public.

11 In contrast to European cities (and even most American resort communities), most US cities have not set up visitors' information centres. New York did not establish its first centre (in Times Square) until the 1980s. Harlem now also has an information centre run by NYC & Company and funded by the federal Empowerment Zone.

12 Only yellow cabs may pick up passengers that hail them on the street. Radio or 'gypsy' cabs must be ordered by telephone.

13 Interview with the director of the Place Matters project for City Lore, 7 October 2003.

14 Information on Fort Greene was provided by Stacey Sutton.

References

Amin, A. (1994) *Post-Fordism*, Oxford: Blackwell.

Debord, G. (1994) *The Society of the Spectacle*, New York: Zone.

Fainstein, S.S. and Fainstein, N. (1989) 'Governing regimes and the political economy of redevelopment in New York City', in J.H. Mollenkopf, T. Bender and I. Katznelson (eds) *Power, Culture and Place: essays on the history of New York City*, New York: Russell Sage.

Fainstein, S.S. and Gladstone, D. (1999) 'Evaluating urban tourism', in D.R. Judd and S.S. Fainstein (eds) *The Tourist City*, New Haven: Yale University Press.

Geddes, M. (2000) 'Tackling social exclusion in the European Union? The limits to the new orthodoxy of local partnership', *International Journal of Urban and Regional Research*, 24 (December): 782-801.

Harvey, D. (1990) *The Condition of Postmodernity*, Oxford: Blackwell.

Hoffman, L.M. (2003) 'Revalorizing the inner-city: tourism and regulation in Harlem', in L.M. Hoffman, S.S Fainstein and D.R. Judd (eds) *Cities and Visitors: regulating people, markets, and city space*, Oxford: Blackwell.

Hoffman, L.M., Fainstein, S.S. and Judd, D.R. (eds) (2003) *Cities and Visitors: regulating people, markets, and city space*, Oxford: Blackwell.

Judd, D.R. (2003) 'Visitors and the spatial ecology of the city', in L.M. Hoffman, S.S. Fainstein and D.R. Judd (eds) *Cities and Visitors: regulating people, markets, and city space*, Oxford: Blackwell.

Lauria, M. (ed.) (1997) *Reconstructing Urban Regime Theory*, Thousand Oaks CA: Sage.

MacCannell, D. (1976) *The Tourist, a new theory of the leisure class*, New York: Schocken.

Markusen, A.R. and S.S. Fainstein (1993) 'Urban policy: bridging the social and economic development gap', *University of North Carolina Law Review*, 71: 1463–86.

New York Department of City Planning (2004) 'Change in population 1990–2000', available http://www.nyc.gov/html/dcp/pdf/census/pl4b.pdf (accessed 6 December 2004).

Sutton, S. (2002) 'Making an enclave: identity, entrepreneurship and place attachment', unpublished paper, Department of Sociology, Rutgers University; available <http://sociology.rutgers.edu/colloquium/pdf/stacey_sutton.pdf> (accessed 20 January 2005).

Wilson, W.J. (1980) *The Declining Significance of Race*, Chicago: University of Chicago Press.

Wilson, W.J. (1990) *The Truly Disadvantaged*, Chicago: University of Chicago Press.

9 Selling Miami

Tourism promotion and immigrant neighbourhoods in the capital of Latin America

Gastón Alonso

Miami is not like any other place. Miami is America's only national playground city in the only US tropics – built by Americans for all Americans – for people like you.... Who did it? "We, the people" did it. America built Miami because Americans wanted Miami.

(Miami Chamber of Commerce 1940)

The developers and hoteliers who built Miami at the turn of the twentieth century attracted new residents and tourists by mixing faux 'Spanish' architecture and lush tropical vegetation to create a tropical*ized* (Aparicio and Chávez-Silverman 1997) 'playground city'. It was a place imbued with traits and images of sensuality, heat and foreignness that are linked in the popular imaginary with the Tropics and are associated in Western thought with notions of exotic Otherness. By the end of the century, their city had become a regional trade and financial centre as well as a gateway for immigrants from the Caribbean and Latin America. Ironically, despite Miami's 'Latin Americanization', to borrow Mike Davis's (2000) phrase, tourism boosters today are selling a *less* tropicalized image of the city than the one sold earlier in the twentieth century. Rather than highlighting Miami's Caribbean and Latino neighbourhoods and cultures, they emphasize its 'cosmopolitan sophistication', modern downtown skyline, restored Art Deco hotels and world-class shopping centres. In fact, the working-class immigrants who give the city its contemporary 'tropical rhythm' – and on whose labour the tourism industry depends – are mostly left out of the image marketed to tourists.

This is an interesting choice on the part of local boosters, since city governments and tourism bureaus around the world are increasingly seeking to attract tourists precisely by showcasing their cities' immigrant neighbourhoods and cultures as 'exotic' places for tourists to experience. In the face of deindustrialization, cities as diverse as Amsterdam, Birmingham, Boston, Lisbon, Melbourne, New Orleans, Toronto and Vancouver have turned to tourism as a strategy of local economic development (Hoffman *et al.* 2003; Judd and Fainstein 1999). In seeking to distinguish themselves from other urban areas, these cities have promoted immigrant neighbourhoods as destinations where tourists can eat 'ethnic' foods, buy 'ethnic' goods and experience 'ethnic' cultures (Hoffman 2003; Lin 1998; Rath 2005). As Laguerre (1999: 99) has pointed out, these Chinatowns and Koreatowns, Little Italys and Little Bombays constitute 'minoritized spaces' that emerge 'as a result of segregated practices that bar a group from participating on an equal footing in the mainstream affairs of state'. As such, these neighbourhoods are distinguished by their populations, commerce and social life, and sometimes by their architecture. The drive to attract tourists has led city officials to see the value of turning these often neglected spaces into aesthetized spaces of consumption and entertainment, showcased in promotional campaigns and guidebooks (Zukin 1998). In this way, immigrant cultures are providing cities with what Rath (2005) has referred to as a 'diversity dividend' or an 'ethnic advantage'. At the same time, tourism industries can provide immigrants with economic opportunities – and not just as high- or low-skilled service workers, but also, importantly, as small entrepreneurs and producers of tourist attractions. By working with city officials to transform their neighbourhoods into 'ethnic destinations', immigrant entrepreneurs can tap into the transnational flows of capital, goods and people that are part of tourism economies. Thus, as Rath suggests, 'expressions of immigrant culture can be transformed into vehicles for socioeconomic development to the advantage of both immigrants and the city at large'.

In the light of these developments, the present chapter asks: Why have immigrant entrepreneurs in Miami *not* been able to translate their 'ethnic difference' into an 'economic advantage' that could provide them entrance into the tourism industry? Answering these questions will help identify circumstances under which 'ethnic tourism' can be deployed more effectively as a strategy of socioeconomic development. I begin by setting the scene as it was, tracing the development of Miami from a Southern resort town to a multiracial global city. I examine the place that Little Havana, the city's most prominent immigrant neighbourhood, occupies today within Miami's socioeconomic landscape. I then turn my focus on the promotional campaigns mounted by the Greater Miami Convention and Visitors Bureau (GMCVB). As part of the city's 'critical infrastructure' (Zukin 1991), this organization plays a major role in the valorization of neighbourhoods. In the concluding section, I argue that attempts by immigrant entrepreneurs to transform Little Havana into an 'ethnic tourist destination' have been undermined by the lack of distinctive value attached to Cuban ethnicity within the region's pervasively 'Latin Americanized' landscape.

Rath has identified a number of circumstances that might 'gum up the transformation of cultural resources into an economic asset'. Among these are class- and gender-based conflicts between immigrant entrepreneurs in the tourism industry and their co-ethnic employees, resentments from immigrants at the transformation of their neighbourhoods into tourist attractions, the 'homogenization and fossilization' of neighbourhoods by increased government regulation, and tensions between tourists and locals regarding the use of public space (Rath 2005). The case study of Miami presented here augments Rath's list by suggesting that the value attached to specific ethnicities within particular symbolic contexts can also 'gum up' efforts by immigrant communities to use 'ethnic tourism' as a strategy of socioeconomic development.

The 'Magic City' becomes the 'Capital of Latin America'

During most of the nineteenth century, South Florida was an uninviting place populated mostly by alligators and mosquitoes. Rail transportation connecting it to the rest of the country arrived at the close of the century through the efforts of Julia Tuttle, a widow who owned land north of the Miami River, and of Henry M. Flagler, a partner of John D. Rockefeller who had come to Florida in 1883 to build a chain of luxury hotels connected together by a railroad. Flagler's hotels in St Augustine and Palm Beach catered to wealthy North-easterners seeking warmth, leisure and pleasure during the winter months. Tuttle persuaded Flagler to extend his railroad all the way to the Miami River and to build a hotel at the mouth of the river. In 1896 Flagler's railroad arrived, his Royal Palm Hotel opened, and the City of Miami was incorporated (Akin 1992; Martin 1949).

The railroad fuelled the local economy, as land speculators and developers descended on the region. Promoters like Flagler, Carl Fisher, the founder of the resort town of Miami Beach, and George Merrick, founder of the exclusive community of Coral Gables, spent millions of dollars in advertising campaigns to sell the new city as a tropicalized playground. A typical promotional brochure of the early 1900s read: 'Miami welcomes you with the song of the tropics. Leave winter behind, fling care to the icy winds, come to Miami and play at being eternally young again' (Muir 1953: 144). To maximize the 'tropical lure' of what they dubbed the 'Magic City', developers built a backdrop of Mediterranean Revival architecture, complete with red tile roofs, courtyard fountains and arched passageways. Sun worshippers arrived in droves to relax during the winter months, or even to settle for good and live out the romantic lifestyle promised by the city's promoters. As a result, between 1910 and 1930 the population of Miami exploded from 4955 to 142,955 (US Census Bureau 1995). Left out of the promotional campaigns used to sell this 'imaginary landscape' (Zukin 1991) was the city's black community. Native-born blacks and West Indian immigrants drained and reclaimed the Everglades for agricultural production and cleared native mangroves for building the hotels and housing subdivisions that made Northern entrepreneurs wealthy. Despite its reputation as a tourist destination populated by transplanted Northerners, Miami was actually a segregated Southern town, and black labourers were confined to neighbourhoods with practically non-existent municipal services and substandard housing. Moreover, the local police department and the Ku Klux Klan made sure that blacks were kept in the shadows of the thriving 'Magic City' that the Northern whites were coming to enjoy (Dunn 1997; George 1978; Mohl 1987).

World War II transformed Miami, laying the foundations for its emergence as a trade and financial centre. During the war, the military set up a training command centre and took over about 85 per cent of the hotel space on the beach. After the war, it continued to operate airfields in the region, pumping millions of dollars into the local economy, and the Army Corps of Engineers drained large segments of the Everglades, opening up thousands of acres to new housing developments and spurring the local construction industry. At this time, Miami became a major hub for the airline industry, as planes replaced trains as tourists' preferred means of transportation. Miami also became a centre for trade with Latin America, as goods headed to and from the Caribbean and South America passed through its rapidly expanding airport and port facilities. Attracted by the warm climate, sunny beaches and jobs in the region's growing airline, tourism and trade industries, and drawn by improvements in pesticides and air conditioning that enhanced the quality of life in the area, thousands of white Northerners, many of them former enlisted men who got 'sand in their shoes' during the war, headed to South Florida (Mohl 1990; Mohl and Mormino 1995; Mormino 1997). As a result, the population of Miami surged from 267,739 in 1940 to 495,084 in 1950 (US Census Bureau 1995).

Jews comprised a large segment of newcomers to the area, and the Jewish population of the county increased from a mere 5500 in 1940 to 55,000 in 1950. Fisher had placed restrictive covenants on land parcels that he sold and had hung 'No Jews' or 'Gentiles Only' signs on the doors of hotels that he built. After the war, however, Jewish investors from New York led a building boom on Miami Beach. To attract fellow Jews they built luxury hotels and high-rise apartment buildings. As Jews opened synagogues and Hebrew schools, delicatessens and kosher markets, Fisher's resort town became known as 'a Southern borscht town' (Moore 1994: 35). In the process, the newcomers replaced the theme-park Mediterranean architectural style favoured by the entrepreneurs of the early 1900s with a distinctive 'Miami Modern' (MiMo) style. The city developed a less tropicalized ambiance.

The changes that Miami underwent during the 1950s pale in comparison to the massive demographic and economic changes it underwent during the decades to follow. The Cuban Revolution set in motion forces that brought hundreds of thousands of Cuban exiles to Miami, turning the city into a bilingual and bicultural place. In the words of Joaquin Blaya, the general manager of one of the city's Spanish-language television stations,

The world the Cubans created made it comfortable for other Latins. Many here think Miami is the best city in all of Latin America, the status capital of South America. If you're Venezuelan or Colombian and you have not been to Miami, you are not 'in', you are 'out'.

(Levine 1985: 61)

Thus, during the 1970s, Latin Americans began travelling to Miami as tourists seeking a taste of US mass culture and as shoppers seeking luxury goods not available in their home countries. Latin Americans also came as investors, depositing money in local banks and buying apartments in luxury condominiums in Key Biscayne or along Brickell Avenue, or homes in new subdivisions located in the western edges of the county (Mohl 1990). During the 1980s and 1990s, the economic and political upheavals and dislocations occasioned by the integration of the hemisphere's economy led thousands of people from the Caribbean Basin and Latin America to leave their countries in search of better opportunities. Many were drawn to Miami by the large and visible presence of other immigrants in the city and by the availability of jobs in the region's expanding high- and low-skill service sectors. In particular, the informalization of labour practices in Miami's low-skill garment, construction and service sectors (including tourism) facilitated the incorporation of large numbers of undocumented immigrants. By 2000 Miami–Dade County[1] had the largest concentration of foreign-born residents of any urban county in the country, with Latinos making up 57 per cent of the county's 2,253,362 residents, while non-Latino whites and blacks comprised only 20 per cent each (US Census Bureau 2002). Cubans, however, no longer constituted the majority of the county's Latino population, and 30 per cent of the black residents were born abroad, predominantly in the Caribbean nations of Haiti, Jamaica and the Bahamas (Metro–Dade County Planning Department 1995). In the process, as Allman (1987: 149) has explained in his book on Miami, the city became much less tropicalized in appearance.

> The reason for this increasingly Latin city's increasingly un-Latin appearance is simple enough: the Latin American fantasy of the Florida good life isn't of red tile, tinkling fountains and Spanish moss, but of high-rise condominiums, chlorinated swimming pools and fast American cars.

As its population became Latin-Americanized, Miami emerged as the trade and financial centre of the Caribbean Basin. Global cities literature emphasizes the role that economic and geopolitical forces played in the integration of Miami's economy with transnational flows of goods and capital (Grosfoguel 1995; Sassen and Portes 1993). However, at the local level, Miami's 'globalization' was orchestrated by Anglo and Cuban 'place entrepreneurs',[2] who reoriented the local economy away from the North-eastern United States and towards the Caribbean and Latin America. This pro-growth coalition promoted the idea that Miami's geographic location and its multicultural/multilingual population rendered it the 'Capital of Latin America'. To attract capital, they pushed local governments to develop the region's transportation and communication infrastructure, enact corporate-friendly labour laws, lower tax rates and provide tax incentives for businesses. In this context, the chambers of commerce and public–private partnerships that act as the local architects of globalization were very successful in attracting multinational corporate headquarters and international financial institutions.

While underemphasized in the literature on global cities, tourism-related industries – transnational hotel chains, cruise lines and airlines – have been central in the articulation of Miami's economy with transnational flows. The leisure and hospitality industries still remain important parts of the local economy, employing 131,000 persons in 2002. The same year, the county hosted 10,231,400 overnight visitors, who pumped a total of $18.5 billion (€20 billion at the time) into the local economy and generated $84 million in tourism taxes and $595 million in state sales taxes. While Miami remains a tourist mecca, it now attracts visitors from around the world. In 2002, domestic visitors, mostly from North-eastern states, made up only 52 per cent of all overnight visitors, while international visitors, largely from Latin America, accounted for 48 per cent (GMCVB 2003a). Tourism-related industries in the region thrive on the cheap labour of low-skill immigrants. In the city's racialized division of labour, management positions in the hospitality and leisure industries tend to be overwhelmingly occupied by native-born Anglos, front-of-the-house jobs such as reception desk clerks and waiters by Latino immigrants, and back-of-the-house jobs like hotel housekeeping and food preparation by black immigrants. These industries are notorious for their informalized labour practices, including paying workers below the minimum wage, making illegal payroll deductions and violating health and safety laws (Stepick *et al.* 1994).

La Pequeña Habana in globalized Miami

Within the landscape of globalized Miami, Little Havana has traditionally served as a 'gateway' for working-class immigrants. The community has its roots in the Cuban migration of the 1960s. The first wave of exiles was composed of political elites associated with the Batista regime, upper-class landowners, industrialists and employees of US companies on the island. By October 1962, approximately 248,000 Cubans had sought refuge in the United States, most remaining in Miami. This wave of migration ended that month, when tensions over the Cuban Missile Crisis stopped direct flights between the island and the United States. A second migratory wave began in 1965, when the US agreed to sponsor flights from Cuba to South Florida. For eight years, these 'freedom flights' brought 297,318 Cubans to Miami. Unlike earlier exiles, these immigrants were mostly drawn from the urban working class (Masud-Piloto 1996; Pedraza-Bailey 1985).

The migration of Cubans to Miami occurred within the context of Cold War geopolitics, which led the United States to interpret the tightening alliance between Castro and the USSR as a major threat to its economic, geopolitical and military control of the hemisphere. In order to destabilize the Castro regime, the US encouraged disaffected Cubans to leave the island. The image of thousands of exiles arriving in South Florida was used on the world stage to discredit the Castro regime and to legitimize US policies towards it. Meanwhile, the migration of Cuba's best qualified professionals undermined Castro's efforts to restructure the island's economy. In order to encourage and facilitate this mass exodus, the US not only sponsored 'freedom flights' but also granted Cubans legal entry into the country. Moreover, it made programmes of settlement assistance available to the exiles which were seldom offered to other immigrant groups. The Cuban Refugee Program (CRP) provided exiles with direct cash assistance and small business loans, as well as offering bilingual education, professional recertification and job-training programmes. From 1961 to 1972, the federal government spent over $1 billion on this programme, aiding 75 per cent of all exiles (Mohl 1990). Furthermore, during the same decades the CIA spent over $100 million dollars a year in the local economy, placed up to 20,000 exiles on its payroll and funded a number of 'front' operations – measures which all provided start-up capital for Cuban-owned businesses (Argüelles 1982; Torres 1999).

Little Havana emerged in this context. Early exiles were attracted to the area south-west of the central business district for a number of reasons. First, Riverside (as the neighbourhood was then known) was already home to a small community of Cubans who had arrived around 1933 following the overthrow of president-turned-dictator Gerardo Machado. Second, while Riverside had been an economically vibrant, predominantly white neighbourhood during the first half of the century, by the 1950s it had become economically depressed, and many residents and business owners had abandoned the area. The decrepit houses and storefronts they left behind provided Cubans with low-rent housing and commercial spaces. Last, the neighbourhood afforded easy access to the Cuban Refugee Program and to jobs in the downtown district. Cubans took advantage of resources provided by the programme, as well as receiving 'character loans' from fellow exiles, and they were soon opening small businesses along the area's main commercial artery SW 8th Street. By the late 1960s, the high concentration of Cuban residents and the increasing presence of Cuban-owned businesses had given the neighbourhood a decidedly Cuban flavour. Locals began referring to it as La Pequeña Habana, or Little Havana, and to the neighbourhood's commercial artery as la Calle Ocho or 8th Street (Abrahamson 1995; Curtis 1980).

During the next decade, many Cubans began to move up the economic ladder. They continued to draw on settlement assistance and community resources to open small businesses – the backbone of the much-extolled Cuban enclave economy (Portes and Bach 1985; Wilson and Portes 1980) – and, importantly, to gain entrance into the region's rapidly restructuring primary labour market. Cubans of rising socioeconomic status moved into subdivisions built by Cuban developers in the western part of the county. When Latin-American immigration skyrocketed during the next decades, these subdivisions also became the preferred area of settlement for middle- and upper-class immigrants. Meanwhile, Little Havana decayed, as storefronts were boarded up and crime rates grew. The neighbourhood, however, continued to serve as a port-of-entry for recently arrived Latino immigrants in need of low-cost rental housing. In fact, the increasing presence of Central American immigrants in the area has led locals to call the neighbourhood 'Little Managua'.

The neighbourhood is perhaps best known as the home of the annual Calle Ocho Street Festival, launched in 1978 by the Kiwanis of Little Havana, a community service organization made up of Cuban entrepreneurs. In 1982, Miami city officials offered the organizers funds to expand the event into a two-week festival. Today, Carnaval Miami, as the festival is known, includes a music concert, a beauty pageant and an 8K run, and culminates with the street festival on Calle Ocho. Self-billed as 'the largest celebration of Hispanic culture in the United States', the festival attracts a million people – mostly Latino immigrants – and is broadcast throughout the hemisphere. As the neighbourhood has changed and the festival has grown, it has become less a celebration of Cuban exile identity and more a festival of Pan-American cultures with food kiosks and musical acts from throughout the hemisphere. At the same time, large corporations eager to tap into the growing Latino market have displaced local merchants as sponsors. Thus, while this 'carnival of consumption', as a local newspaper dubbed it,[3] has brought international recognition to the neighbourhood, it has failed to substantially benefit local small merchants.

In recent years, a group of neighbourhood entrepreneurs have spearheaded a plan to revitalize Little Havana by capitalizing on the area's internationally recognized name. At the centre of the plan has been the Latin Quarter project funded by the non-profit East Little Havana Community Development Corporation (CDC). Built at the corner of SW 8th Street and 15th Avenue in a faux 'Spanish' architectural style more reminiscent of Miami's early theme-park-style buildings than of Havana, the complex includes a cultural centre, shops, restaurants and 45 condominiums. The plan also involves the restoration of the historic Tower Theater across the street and the creation of a cobblestone walking plaza along 15th Avenue. In 2004, the Little Havana Merchants Association secured funds from the City of Miami to open a tourist information kiosk along Calle Ocho. It is too early to know whether these attempts will attract visitors to the area in ways that will lead to its economic revitalization. Meanwhile, Little Havana continues to be among the poorest neighbourhoods within the poorest city in the country; its deteriorating bungalows and apartment buildings are home to elderly Cuban residents largely dependent on government assistance and to Central American immigrants dependent on the city's low-wage service economy.

Selling 'Tropicool Miami'[4]

Miami's economic growth has always depended on the ability of city boosters to attract residents, investors and, importantly, tourists. As discussed above, early promotional campaigns were orchestrated by the private organizations of developers such as Flagler, Fisher and Merrick, and subsequently by the Miami Chamber of Commerce. Under the leadership of E.G. Sewell, the chamber sponsored nationwide campaigns inviting people to 'Visit Miami: where the summer spends the winter'. Chamber-sponsored newspaper ads, billboards and brochures featured images of beautiful healthy young white women basking in the sun against the backdrop of the region's 'Mediterranean' architecture and tropical fauna. One such ad proclaimed Miami to be 'The wonder city of the tropics', while another ad rhapsodized that where 'once the Seminole and alligators held dominion in limitless stretches of glades, one now finds canals, streets and highways thrusting forth their claim to the nice impenetrable land' (Miami Chamber of Commerce 1930). To ensure good publicity for the 'Magic City', the chamber invited journalists to luxurious trips and staged annual agricultural and art expositions as well as sporting events (Bush 1999; George 1981). Pioneered in Miami, these marketing strategies are now used by tourism boosters around the world.

Today the job of attracting tourists to Miami is mainly the responsibility of the Greater Miami Convention and Visitors Bureau (GMCVB). Founded in 1984, the bureau is a public-private partnership between local governments and over 1100 businesses in the hospitality and leisure industries. To promote Miami, the bureau distributes news releases and feature stories; arranges accommodations, itineraries and sightseeing tours for visiting journalists and writers; offers 'familiarization tours' to travel agents and tour operators from around the world; attends tourism industry tradeshows in the US and abroad; and develops relationships with tourism representatives in select markets around the world.

The promotional strategy deployed by the GMCVB emerged from its attempts to overcome the negative reputation Miami developed during the 1980s. At the start of that decade, the global economic and demographic forces described above had thrown the city into a state of crisis. The arrival of 60,000 Haitian refugees between 1977 and 1980 and 125,000 Cubans during a five-month period in 1980 had strained city resources and exacerbated ethnic tensions. These tensions were manifested by the passage in 1980 of a referendum overturning a 1973 ordinance that had declared the county officially bilingual, and by the outbreak the same year of the largest instance of urban rebellion in the nation's history until that date. The 'Liberty City riot' was a response by the black community to years of police violence as well as to growing feelings of marginalization due to increased job competition from immigrants (Mohl 1990). By this time, Miami had also become the main port-of-entry for illegal drugs, and the city's murder rate was the highest in the nation. In news stories circulated around the globe, Miami became associated with racial conflict, crime and drugs. Images of sunny beaches were eclipsed by images of Cuban refugees living in a tent city under Interstate highway 95, Haitian refugees arriving ashore in homemade rafts, Liberty City smouldering during the riots, and drug shootouts on the streets of Miami. As a consequence, both domestic and international tourism to the area declined. In addition, the advent of inexpensive air travel to Caribbean destinations and the opening of Disney World in Central Florida further eroded tourism to South Florida. After *Time Magazine* published a cover story dubbing Miami 'Paradise Lost' (1981),[5] local tourism boosters founded the GMCVB and launched a campaign to revive the city's image.

For inspiration they turned not to the city's growing immigrant communities – sources of so much local-level tension – but rather to the city's past, and in particular to the Art Deco district at the southern edge of Miami Beach. The Art Deco district first came about after a hurricane destroyed the area in 1926. Rather than replicate Fisher's faux 'Spanish' style, the developers turned to the relatively cost-efficient and increasingly popular Art Deco style. The Tropical Deco style they favoured was characterized by the use of picturesque porthole windows and decorative murals and reliefs of tropical images such as flamingos and palm trees. During the post–World War II years, the district decayed, as tourists flocked north along Collins Avenue to new, larger luxury hotels. Many of the district's independently owned small hotels were either boarded up or converted into single-room occupancy buildings populated mostly by elderly retirees and new immigrants. But the district found a saviour in Barbara Baer Capitan, an interior design magazine editor and transplanted New Yorker. In 1976 she launched the Miami Design Preservation League, and within three years 1200 buildings were listed in the National Register of Historic Places. To attract investors to restore the district's dilapidated buildings, Capitan and interior designer Leonard Horowitz created a new 'signature' style for the district by transforming the original muted white or cream facades of the buildings to a purposely eye-catching palette of hot pink, lavender, peach and lime (Ellwood 2002: 288–93). The revitalization of South Beach – as the district was now redubbed – was furthered by the international popularity of the 1980s television series *Miami Vice*. The police drama with the look of a music video used the city's gleaming downtown office towers, opulent waterfront mansions, and, importantly, its pastel Art Deco buildings as a stylized backdrop to its story lines of drug trafficking and police corruption (Zukin 1991: 242–50). By the end of the decade, the GMCVB sought to exploit the district's revitalization and its new internationally recognized image by inviting fashion photographers to the city. The bureau facilitated accommodations for visiting photographers, organized tours of the city's sights and expedited permits for their shoots. Soon South Beach became recognized as a cheaper alternative to other shooting locations, and the city's pastel buildings and sunny beaches filled the pages of chic fashion magazines. Overnight, Miami went from a 'Paradise Lost' to a fashionable destination.

This promotional strategy is evident in the 224-page glossy *Vacation Planner* published by the GMCVB every year.[6] The guide is distributed to journalists and tour-book writers around the world and is available in over 50,000 hotel rooms in the county. The two most prevalent images in the 2003 edition (GMCVB 2003b) are the downtown high-rises built during the last few decades as regional headquarters for transnational capital and the Art Deco hotels renovated by multinational hotel corporations during the same period. These are the images the bureau promotes as the 'signature' of the Miami-as-commodity which tourists are invited to consume. The guide welcomes visitors by painting the following vision of this city-as-commodity:

> Surrounded by the gentle wells of the Atlantic Ocean, Greater Miami takes pride in its turquoise waters, sunny skies and year-round warm temperatures. In addition to our natural beauty, a myriad of attractions, recreational activities, museums, festivals and fairs add a spicy sense of fun. World-class cuisine, sizzling nightlife and infinite shopping opportunities are just part of the magical appeal of the community.
>
> (GMCVB 2003b: 19)

In full-page spreads, the guide touts the 'sights, shopping, dining and nightlife' of the select neighbourhoods of Miami Beach, South Beach, Downtown, Coconut Grove, Bal Harbour, Coral Gables, North Beaches, Key Biscayne, North-east Miami–Dade and South Miami–Dade including the Everglades, 'the largest wilderness in the Eastern United States' (GMCVB 2003b: 33). The visitor is informed that the Art Deco district is 'recognized globally as one of Greater Miami and the Beaches' unique attractions' (GMCVB 2003b: 24). The accompanying pictures of tanned and scantly clad beautiful people strolling down Ocean Drive or dancing at the district's stylish nightclubs let the visitor know that, as 'a lush, tropical environment', the Miami area

> evokes one great movie set, a fantasy world peopled with models and celebrities. But on this stage everyone can enjoy the same alluring

lifestyle, whether relaxing poolside or beachside, lunching at umbrella-shaded tables or zipping along Ocean Drive on roller blades.
(GMCVB 2003b: 23)

In the 'fantasy world' marketed by the bureau, the celebrities themselves, in fact, become 'attractions'.

Where else can you salsa, merengue or hip hop the night away in clubs, bars and lounges all over town [the guide asks] while rubbing elbows with stars of entertainment, sports and music industries, as well as throngs of models who walk the catwalks and pose beneath the sunshine's ray by day?
(GMCVB 2003b: 128)

In a section promoting the region's 'shopentertainment', tourists are reassured that

to make the Greater Miami and the Beaches' shopping experience even sweeter you'll encounter knowledgeable, multilingual staff who will provide you with attentive service, whether you are shopping for prestigious designer names or everyday great deals.
(GMCVB 2003b: 64)

To a major extent it has been the Latin-Americanization of the city's population, economy and cultural life that has given Miami its 'tropicool' atmosphere, its 'multilingual' labour force and its status as a global city. The promotion of the city's 'tropicoolness', however, has not emphasized its working-class immigrant neighbourhoods. In fact, the most prominent of these neighbourhoods, Little Havana, plays just a minor role in the vision of Miami-as-commodity marketed by the guide. Unlike upscale neighbourhoods such as Miami Beach, Coral Gables and Coconut Grove, Little Havana is mentioned only briefly. It is included as part of a list of 'Other Neighborhoods' that is provided in the 'Neighborhood Guide' section. In Little Havana, visitors are informed, 'everything is authentic from the fruit stands and cigar factories to the eat-at windows of the cafeterias crowded with patrons passionately discussing politics' (GMCVB 2003b: 27). However, the 14-page list of restaurants that instructs visitors as to where they can taste 'Miami's signature cuisine' only includes four restaurants actually located in Little Havana (GMCVB 2003b: 114).

Not surprisingly, then, a GMCVB-sponsored study found that in 2003 the vast majority of overnight visitors neither stayed in nor visited Little Havana. Some 43 per cent reported staying in hotels on Miami Beach, 16.8 per cent near the airport, 10.5 per cent in the North Dade / Sunny Isles area, 10.5 per cent in Downtown, 3.8 per cent in South Miami–Dade and 5.5 per cent in the upscale neighbourhoods of Coconut Grove, Coral Gables and Key Biscayne. The same study found that while 68.9 per cent visited the Art Deco district, only 5.8 per cent went to Little Havana. Only 18.2 per cent reported liking the city's 'international ambiance', compared to 60.3 per cent who liked the weather and 32.2 per cent the beaches.[7] The city that most tourists experience, then, is a place of beautiful people and warm beaches, of world-class shopping opportunities and Art Deco architecture, and a place where the majority of immigrants seldom become attractions or, more importantly, the producers of attractions.

The relative value of ethnicity in hyper-ethnic contexts

What factors explain the relatively minor role that Little Havana plays in the image of Miami-as-commodity sold to tourists? A partial answer lies in the nature of tourism promotion, as is also suggested by other chapters in this volume. In urban-coastal destinations like Miami, tourism boosters can concentrate on the relatively easy target of selling 'sand and surf', rather than on the much harder task of promoting urban environments. Moreover, since effective marketing campaigns work by disconnecting the representation of commodities from their production, the tourism boosters have little incentive to draw visitors to the poor neighbourhoods that are home to the very people they see labouring in the tourism industry. There is, of course, nothing particularly new about all of this. As observed above, the promotional efforts mounted by Flagler, Fischer and the Miami Chamber of Commerce during the 'Magic City's' early years also did not invite tourists to 'experience' the blighted conditions of Colored Town, even though the tourism industry at the time was dependent on the labour of the black residents of that segregated neighbourhood. Furthermore, the boards of directors and executive committees of tourism bureaus, which are public–private partnerships, tend to be overwhelmingly composed of industry representatives, and most of the places that bureaus tend to designate as 'must see' are in those neighbourhoods where the major hotel chains, shopping centres and established tourist attractions are found.

The other part of the answer moves the analysis from issues of place promotion to issues of place production, and focuses on the nature of the neighbourhood itself. After all, aside from their valorizing role, tourism bureaus are dedicated to the *marketing* of places, not to their production. Little Havana lacks unique or significant natural or human-made 'attractions'. There are a number of memorials to Cuban figures, but no particularly significant historical sites, museums or parks designed for tourists. Moreover, during its decay in the 1980s and 1990s, the neighbourhood had a relatively high crime rate, and there is no good transportation infrastructure between it and the areas where most hotels are located. Thus, there is currently very little that renders Little Havana an attractive 'destination' for outsiders. Of course, places are produced by public policy choices about land use, zoning regulations, infrastructure development and allocation of public resources. As other cities have decided to turn their immigrant neighbourhoods into 'destinations', they have poured funds into developing tourist-oriented attractions in these neighbourhoods, making them easily accessible and reducing their crime rates. The question, then, becomes: Why have immigrant entrepreneurs in Little Havana not been able to trade on their 'ethnic difference' to transform their neighbourhood into a marketable 'destination'?

To try to get 'on the tourist map', the Little Havana Merchants Association placed a special advertising section in the 2004 edition of the GMCVB's *Travel Planner*. The eight-page spread proclaims that 'Little Havana is a cultural destination' and invites tourists to walk down Calle Ocho and

> observe cigars being hand-rolled, dine at cafeterias and restaurants serving Cuban sandwiches and café con leche, drink fresh coconut and sugar cane juice, dance to Latin music blaring from music stores, buy Little Havana souvenirs and Cuban memorabilia and hear the locals passionately discussing politics or playing dominoes.

Among the businesses advertised in the section are 'the authentic Spanish tavern' Casa Pansa, El Pub Restaurante where tourists can 'enjoy Cuban entrees and desserts clipped from cookbooks from the 1900s', the area's 'official souvenir store' Little Havana To Go, and El Credito Cigar Factory where tourists can 'feel what's real'.

The construction of the 'Spanish'-looking Latin Quarter project and this advertising campaign reflect attempts at self-tropicalization – imbuing the neighbourhood and its population with exoticizing mainstream ideas about the 'authentic' Tropics ('Spanish' architecture, 'exotic' foods, 'loud' music, 'passionate' discourse) in order to attract outsiders interested in consuming 'a bit of the Other'. It remains to be seen, though, whether the commercialized expression of the neighbourhood's ethnic identity will garner the attention of tourists and tourism boosters. In cities like the Hague and Birmingham, Chinese and South Asian immigrants respectively have been able to take advantage of their ethnicity to attract public attention and governmental funds. However, there are profound differences between the place that the Hague's Chinatown and Birmingham's Balti Quarter occupy within their respective local landscapes and the position that Little Havana occupies in the Miami landscape. Little Havana is situated in a county where 50 per cent of the population is foreign-born and 68 per cent speaks a language other than English at home (US Census Bureau 2002), and which has undergone a process of 'reverse acculturation' that has Latin-Americanized the mainstream (Stepick *et al.* 2003). It cannot, therefore, easily claim to be more 'ethnic' than other neighbourhoods. In fact, contrary to Abrahamson's (1995: 87) assertion that Little Havana is 'in many respects, a copy of Havana in Miami' and Laguerre's (1999: 101) claim that 'Little Havana is distinguished from the rest of the city because of the visibility of its population and distinctiveness of its street life', the neighbourhood actually neither echoes the actual Havana nor diverges in any remarkable way from the other predominantly Latino neighbourhoods that constitute the bulk of the county today. Even though the commercial life of Little Havana, from its business signage to the conversations on the streets, is conducted primarily in Spanish, and the neighbourhood is replete with Latin American markets, restaurants and pharmacies, the same is the case throughout Miami–Dade County. Thus, in South Florida's hyper-ethnic context, attempts to transform expressions of Cuban ethnicity into an 'economic advantage' are limited by the relatively small 'difference' between those expressions and the rest of this thoroughly Latin-Americanized city.

Not all tourists or locals might even be interested in venturing to Little Havana to consume 'a bit of the Cuban Other'. On the one hand, as Allman (1987) has pointed out, what brings Latin American tourists to Miami – in 2003 they were 63.3 per cent of international visitors (Synovate Miami 2004) – are not fantasies about tropicalized Others, but rather dreams about the North American mainstream culture that is also available for consumption in Miami. It is the 'cool' part of 'tropicool Miami' that they seek to consume, and it is therefore not clear that the merchant association's tropicalizing campaign will appeal to them. On the other hand, tourists who do wish to consume tropicalized expressions of Cuban culture can already do so in the city's hotel districts, without having to venture into the economically depressed and poorly accessible Little Havana area. In short, in the Latin-Americanized landscape of this playground city, there is very little that is unique or different about the commercialized expressions of Cuban ethnicity associated with Little Havana. The rapid demographic transformation of the city has left Little Havana's entrepreneurs with little 'ethnic advantage' that they can capitalize on as they try to insert the community into the transnational flows of capital, goods and people in tourism economies. The case of Little Havana thus shows some of the contextual limitations on 'ethnic tourism' as a strategy of immigrant incorporation.

Notes

1 The terms 'Miami' and 'Miami–Dade County' are intended to be synonymous. The county is composed of 31 municipalities and a large unincorporated territory. The 'City of Miami' refers to the largest municipality.
2 Logan and Molotch (1987: 29) use this term to refer to 'people directly involved in the exchange of places and collections of rent'.
3 C.F. Delgado, 'Carnival of consumption', *Miami New Times*, 7 March 2002.
4 This expression is taken from an address of the GMCVB's official website: www.tropicoolmiami.com (accessed 23 February 2005).
5 J. Kelly, 'Trouble in paradise', *Time Magazine*, 23 November 1981.
6 The GMCVB also publishes a *Delegate Guide* distributed to convention and meeting delegates, a *Travel Planner* sent to travel and trade professionals and a *Meeting Planner* sent to convention and trade show planners. It produces specialized guides

too, such as a gay and lesbian guide portraying Miami as 'the Gay Riviera', a cultural guide, a Spanish-language travel guide and a calendar of events.
7 The study was based on 5400 random interviews conducted during 2003 among overnight visitors to the county (Synovate Miami 2004).

References

Abrahamson, M. (1995) *Urban Enclaves: identity and place in America*, New York: St Martin's Press.

Akin, E. (1992) *Flagler: Rockefeller partner and Florida baron*, Tampa: Florida Atlantic University Press.

Allman, T.D. (1987) *Miami: city of the future*, New York: Atlantic Monthly Press.

Aparicio, F.R. and Chávez-Silverman, S. (eds) (1997) *Tropicalizations: transcultural representations of Latinidad*, Hanover NH: University Press of New England.

Argüelles, L. (1982) 'Cuban Miami: the roots, development and everyday life of an emigré enclave in the US national security state', *Contemporary Marxism*, 5: 27–43.

Bush, G. (1999) '"Playground of the USA": Miami and the promotion of spectacle', *Pacific Historical Review*, 68(2): 153–72.

Curtis, J.R. (1980) 'Miami's Little Havana: yard shrines, cult religion and landscape', *Journal of Cultural Geography*, 1(1): 1–15.

Davis, M. (2000) *Magical Urbanism: Latinos reinvent the US big city*, London: Verso.

Dunn, M. (1997) *Black Miami in the Twentieth Century*, Gainesville: University Press of Florida.

Ellwood, M. (2002) *The Rough Guide to Miami*, London: Rough Guides.

George, P.S. (1978) 'Colored town: Miami's black community, 1896–1930', *Florida Historical Quarterly*, 56: 432–47.

George, P.S. (1981) 'Passage to a new Eden', *Florida Historical Quarterly*, 59: 440–63.

GMCVB (Greater Miami Convention and Visitors Bureau) (2003a) 'Twenty-one reasons why the visitor industry is the number one industry in Greater Miami and the Beaches in 2002'. Available http://www.gmcvb.com/pictures/HotelOccupancys/7J2H1Z4A1H4T3L4B2G3O4T2Y.pdf (accessed 1 March 2005).

GMCVB (Greater Miami Convention and Visitors Bureau) (2003b) *Greater Miami and the Beaches Vacation Planner*, Miami: HCP/Aboard Publishing.

Grosfoguel, R. (1995) 'Global logistics in the Caribbean city system: the case of Miami', in P.L. Knox and P.J. Taylor (eds) *World Cities in a World System*, Cambridge: Cambridge University Press.

Hoffman, L.M. (2003) 'Revalorizing the inner-city: tourism and regulation in Harlem', in L.M. Hoffman, S.S Fainstein and D.R. Judd (eds) *Cities and Visitors: regulating people, markets, and city space*, Oxford: Blackwell.

Hoffman, L.M., Fainstein, S.S. and Judd, D.R. (eds) (2003) *Cities and Visitors: regulating people, markets, and city space*, Oxford: Blackwell.

Judd, D.R. and Fainstein, S.S. (eds) (1999) *The Tourist City*, New Haven: Yale University Press.

Laguerre, M. (1999) *Minoritized Space: an inquiry into the spatial order of things*, Berkeley: Institute of Governmental Studies.

Levine, L.B. (1985) 'The capital of Latin America', *Wilson Quarterly*, 9(4): 46–69.

Lin, J. (1998) *Reconstructing Chinatown: ethnic enclave, global change*, Minneapolis: University of Minnesota Press.

Logan, J.R. and Molotch, H.L. (1987) *Urban Fortunes: the political economy of place*, Berkeley: University of California Press.

Martin, S.W. (1949) *Florida's Flagler*, Athens GA: University of Georgia Press.

Masud-Piloto, F. (1996) *From Welcomed Exiles to Illegal Immigrants: Cuban migration to the US, 1959-1995*, Lanham MD: Rowman and Littlefield.

Metro–Dade County Planning Department (1995) *An Analysis of the Income Circulation in the Black Community*, Miami: Dade County Planning Department.

Miami Chamber of Commerce (1930) (cited in Léon 1997) *Miami by the Sea: the land of palms and sunshine*, 1st edn 1919, Miami: South Florida Historical Museum.

Miami Chamber of Commerce (1940) *America's Miami*, Miami: South Florida Historical Museum.

Mohl, R. (1987) 'Black immigrants: Bahamians in early twentieth-century Miami', *Florida Historical Quarterly*, 65: 171–97.

Mohl, R. (1990) 'On edge: Blacks and Hispanics in metropolitan Miami since 1959', *Florida Historical Quarterly*, 69: 37–56.

Mohl, R. and Mormino, G. (1995) 'The big change in the Sunshine State: a social history of modern Florida', in M. Gannon (ed.) *The New History of Florida*, Gainesville: University Press of Florida.

Moore, D.D. (1994) *To the Golden Cities: pursuing the American Jewish dream in Miami and LA*, New York: Free Press.

Mormino, G. (1997) 'Midas returns: Miami goes to war, 1941–1945', *Tequesta*, 57: 5–52.

Muir, H. (1953) *Miami USA*, New York: Henry Holt.

Pedraza-Bailey, S. (1985) *Political and Economic Migrants in America: Cubans and Mexicans*, Austin: University of Texas Press.

Portes, A. and Bach, R.L. (1985) *Latin Journey: Cuban and Mexican immigrants in the United States*, Berkeley: University of California Press.

Rath, J. (2005) 'Feeding the festive city: immigrant entrepreneurs and tourist industry', in E. Guild and J. van Selm (eds) *International Migration and Security: opportunities and challenges*, London and New York: Routledge.

Sassen, S. and Portes, A. (1993) 'Miami: a new global city?' *Contemporary Sociology*, 22(4): 471–7.

Stepick, A., Grenier, G., Hafidh, H. and Chafee, S. (1994) 'The view from the back of the house: restaurants and hotels in Miami', in L. Lamphere, A. Stepick and G. Grenier (eds) *Newcomers in the Workplace: immigrants and the restructuring of the US economy*, Philadelphia: Temple University Press.

Stepick, A., Grenier, G., Castro, M. and Dunn, M. (2003) *This Land is Our Land: immigrants and power in Miami*, Berkeley: University of California Press.

Synovate Miami (2004) *Visitor Profile and Tourism Impact: 2003 annual report*. Miami: Synovate.

Torres, M. (1999) *In the Land of Mirrors: Cuban exile politics in the United States*, Ann Arbor: University of Michigan Press.

US Census Bureau (1995) *Population of Counties by Decennial Census: 1900 to 1990*, compiled and edited by R.L. Forstall, Washington DC: US Government Printing Office.

US Census Bureau (2002) *Florida Quick Facts*. Washington DC: US Census Bureau.

Wilson, K.L. and Portes, A. (1980) 'Immigrant enclaves: an analysis of the labor market experiences of Cubans in Miami', *American Journal of Sociology*, 86: 295–319.

Zukin, S. (1991) *Landscapes of Power: from Detroit to Disney World*, Berkeley: University of California Press.

Zukin, S. (1998) 'Urban lifestyle: diversity and standardization in spaces of consumption', *Urban Studies*, 35: 825–39.

10 Building a market of ethnic references

Activism and diversity in multicultural settings in Lisbon[1]

M. Margarida Marques and
Francisco Lima da Costa

Whether as wage labourers, restaurant owners, or music and dance performers, migrants are now part of the natural appearance of cosmopolitan global cities. The 'world in one city' (Collins and Castillo 1998) is a result of globalization both 'from above' and 'from below' (Guarnizo and Smith 1998; see also Sassen and Roost 1999). But the integration of migrants into cosmopolitan cities is not a simple or uniform process. Research has repeatedly shown how national frameworks stand in the way of immigrant incorporation (Brubaker 1992) and how local policies can play crucial roles in negotiating the status of migrants (Ireland 1994).

Although much has been said and written about economic integration, it is still not clear how immigrants integrate into the *cultural* dimension of city life, nor how such integration becomes more manifest in the public sphere, notably in the building of an ethnotourist sector. This is the topic we will address here. Our exploration of the Lisbon case may generate knowledge that enables comparative research with other urban contexts. Although mass immigration in Southern European countries is a more recent phenomenon than in the North, Lisbon has long claimed the status of European gateway city. The centuries of Portuguese overseas expansion serve as evidence for that claim, and so do the recent efforts to 'imagine' a worldwide lusophone 'community' (in the terms of Anderson 1991).[2]

This chapter contributes to the debate on immigrant integration by exploring some essential processes in the building of a market of *ethnic references*. It examines the effects that the emergence of such a market may have in the socioeconomic sphere, and it analyses the cultural and political negotiations involved. Portugal forms an interesting case, because the prevailing ideology of exceptionalism, which assumes a racially blind society, is still a potent influence in Portuguese society, cutting across party-political lines (for further elaboration see Boxer 1963 and Castelo 1999; for a critique see Cabral 2001). The exceptionalism thesis is founded on a claim that Portugal occupies a special in-between position in the world system – marginalized by the powerful actors of the centre, while itself marginalizing the downtrodden of the periphery (Santos 2001). The theory holds that Portugal's position has given it a singular approach to Otherness. Notwithstanding this, however, the concomitant idea of inclusive nationhood has given rise to contrasting reactions on the arrival of the Portuguese 'returnees' from colonial Africa in the 1970s, on the later arrival of 'immigrants', and during the campaign for 'Europeanness' that accompanied Portugal's European Union accession in the mid-1980s.

We are particularly interested here in the role played by 'ethnoculture' and 'ethnopolitical entrepreneurs' (Brubaker 2002) in negotiating social inclusion, and in identifying the foreseeable outcomes of this process in both economic and political terms. How does immigrant cultural production come to be seen as respectable and desirable? How does the top-down 'imagining' of a 'singular' Portuguese community rhyme with the bottom-up cultural expression of migrants? How can immigrants come to be seen as an element of enrichment rather than conflict? Who can gain from such a cognitive shift? And what things can be gained?

In a cognitive-analytical framework (Brubaker 2002, 2004; DiMaggio 1997), we explore the opportunities to move from the negative stereotyping that often accompanies the construction of difference in a migration context to a positive mode of stereotyping in an urban leisure setting. Focusing on the role of 'ethnoentrepreneurs',[3] we hope to discover the path from closure and ghettoization[4] to opening and acceptance.

The national context

The Lisbon World Exposition of 1998 was intended to celebrate the turning pages of history – to combine the proudly inherited eight-centuries-old traditions of the Portuguese capital with the high hopes of a modern Lisbon resolutely embarking on the twenty-first century. A public project involving vast financial resources and political engagement on the part of decision-makers, this top-down initiative thus attempted to use historical heritage as a springboard into modernity (Ferreira 2002; Fortuna and Peixoto 2002).[5]

In a broader sense, too, the supposedly exceptional blend of people and cultures that has arisen out of the Portuguese overseas expansion forms an important leitmotiv in the rhetoric of both decision-makers and mainstream creators of culture in Portugal. The resulting singularity of character is said to endow the Portuguese with a special propensity to mingle with Otherness. The national song genre *fado* is often cited as an expression of that cosmopolitan inheritance – as an original cultural product with significant roots in the African and American wanderings of Portuguese seafarers (for information on fado, see Mariza 2004; *Público* 2004a, 2004b).

The origins of Mariza, the new icon of fado propelled to the status of world music star by a Dutch record firm, are traced back to Mouraria, a historic quarter of Lisbon. But the Mouraria of today is an inner-city district that has been the destination of large immigrant inflows since the 1970s, chiefly 'returnees' from Africa, as in the case of Mariza's family. Yet even though Mariza is being promoted as a symbol of the genuine, original mix of peoples that is said to underpin Portuguese culture, fado has never turned into a magnet for immigrant audiences.

On closer inspection of the upper-middle-class precinct adjacent to Expo '98, we see that the cultural expression of diversity is limited to a narrow set of activities. Moreover, any 'diversity references' used in creating an image for this upcoming neighbourhood make no room for the immigrants who live in the Lisbon of today. Other illustrations could also be given of how mainstream references to Otherness contribute to a singularity matrix that is turned towards the outside world – rather than inwards, to the presence of migrants. This one-sided, and rather elitist, concept of culture seems to be premised on a narrow idea of the nation. As Almeida (2000) has pointed out, cosmopolitanism appears to be understood as the dissemination of Portuguese references far abroad, but with no corresponding adoption of references from the other direction.

Though it would be an exaggeration to speak of 'cultural wars', there is abundant evidence of a growing malaise in parts of society that feel excluded from this one-sided notion of culture. This has been underlined by research done on rap music performed by youth of African parentage (Cidra 2002; Contador 2001; Fradique 2003; de Seabra 2005). Work on post-colonial cultures in France and the UK by Hargreaves (2004) and Hargreaves and MacKinney (1997) has identified different forms of expression used by ethnic minority young people to voice their stance. Music is a favoured medium to express both dissatisfaction and a claim to belong, though it is restricted to limited segments of society (see also Laplantine and Nouss 2001). In an underground but vibrant snowball movement, hip-hop music is now mobilizing many creative youth of immigrant ancestry, and its popularity is spreading (Basu and Werbner 2001; on Portuguese hip-hop, see H2T 2003).

Official Portuguese statistics, which recognize only official nationality as a distinguishing category, tend to obscure the real size of the population of immigrant descent. However, research investigating its demographic impact on Portuguese society has shown that it has figured heavily in the population growth of the past decade, the bulk of which has occurred in the Lisbon metropolitan area (Rosa *et al.* 2004). Although the real significance of the migrant-descent youth has yet to be determined, one thing is sure: in spite of their virtual absence from the formal political sphere, they have become a critical factor in the politics of recognition (Taylor 1994). And they are determined to redress what a growing segment of the population sees as a lack of respect for what these youth have proudly come to define as their own cultural heritage.

Older migrant representatives are making the same point. Whether by organizational or political leaders or by other members of immigrant economic and cultural elites, increasing discontent is being voiced about the meagre recognition for the autonomous expressions of communities from former colonies, and about their relegation to an inferior status in Portuguese society (Marques *et al.* 2003). Despite the prevailing rhetoric in political and academic arenas about the exceptional Portuguese ability to relate to other peoples and cultures, the 'return of the caravels'[6] has proved to be anything but a benign cultural issue (Marques *et al.* 2005).

Yet there are also significant segments of mainstream society that subscribe to or even take part in the new forms of cultural expression, thus helping to promote 'alternative' understandings of Portuguese culture. Some people, mostly 'returnees' from former colonies, welcome such developments to avoid relinquishing their ties to their 'lost Eden' of colonial times. There is also increasing popularity in circles with more cosmopolitan outlooks. Still other adherents take a more militant stance aimed at opposing the normative, top-down monopoly on the cultural expression of nationhood and modernity as claimed by the established cultural elite.

In concert with the growing pursuit of multicultural expression by migrant communities and the wider adherence to the global 'ideology of ethnic diversity' (Fischer 1999), the latter activists have had a significant role in augmenting the present 'ethnoentrepreneurial' supply, rendering it much more visible in the public sphere. The ethnoentrepreneurs occupy a bundle of occupations, and include artists, media professionals, politicians and welfare workers, both in national and transnational spheres. They play a major mediating role, acting as gatekeepers and bridges to the public sphere, and also as marketeers that try to boost demand itself (Crane 1994).

Studies made in France about jazz music and 'Black Paris' (Jules-Rosette 1994; Yeary 2003) have shown how jazz gained roots in post-war France in a counterculture of resistance to the American-oriented mainstream. The idea of closure and resistance was crucial in building a Parisian market of cultural minority references. As closure appears to be a condition for 'authenticity' (van den Berghe 1994; Hargreaves 2004), whilst authenticity appears to be an essential element in the outward marketing of ethnic diversity, we will be particularly keen to study any feedback effects between closure and marketization.

A word of caution should be entered about the use of ethnic labelling. As Brubaker (2002) has argued, the idea of 'ethnic group' often reflects a 'groupist' turn, rooted in some sort of 'folk sociology' – which, we might add, also underpins the basic dyadic relationship of tourism (van den Berghe 1994). At the same time, however, 'us' and 'them' as essentialized constructions can also be politically powerful notions in the construction of ethnicity (Vermeulen 1999). The political dimension is clear in the way activist ethnoentrepreneurs contribute to this ongoing construction of a market of 'exotics at home' (Di Leonardo 1998). They are striving to overcome the subordinated status of minority cultural production and to assert an alternative mode of equitable ethnic relations.

In this chapter, we examine the making of a market of ethnic diversity using the insights we have gained in two case studies. The first involves the Diversity Festival *(Festa da Diversidade)*, an initiative by an anti-racist organization which takes place in the historic neighbourhood of Mouraria, now associated with immigrants of different origins, in Lisbon city centre. The second focuses on the Sabura ethnic tourism project, a grassroots initiative in an outlying slum district of Lisbon, mainly inhabited by African migrants and their offspring. We attempt to identify the necessary preconditions for a market of ethnic diversity, and we argue that, in order for such a market to thrive, a fundamental dilemma has to be resolved: a closure inside certain references is needed to safeguard difference and 'outsidedness' (vis-à-vis the mainstream), whilst at the same time the relations with mainstream institutions are essential, for without them there can be no communication and no bridges, and hence no marketization (see van den Berghe 1994).

Building a setting of ethnic references

Immigrants – who were initially kept hidden away in workplaces and poor districts of town – are now surfacing in the times and spaces of everyday urban life. The ethnic supply side is becoming particularly noticeable in certain districts, and 'ethnic tourism' – still barely acknowledged as an autonomous sphere of commerce – is paving its way into policymaking discourse.[7] The incipient opportunities in urban contexts can be framed in both economic and urban regeneration terms. Although no systematic study has ever been made of the relationship between immigrants and the leisure industry in Portugal, the media have been reporting and reflecting the growing supply of – as well as demand for – certain ethnic references. The High Commission for Immigration and Ethnic Minorities (ACIME), a Portuguese central government body, leans towards a similar celebratory tone, and has a link on its website to a web page on the 'tastes of the world' (ACIME 2005).

In Lisbon, considerable state-led investments have been made in the past two decades to regenerate significant fringes of urban space. Other areas have been left to dereliction. The few remaining immigrant shanty-towns at the city's edge are now under siege from urban developers, and are being tenaciously defended by local residents and institutions. We will compare two urban districts that have become associated with immigrants and cultural diversity. The fact that these manifestations have different origins and follow different patterns – partly by virtue of the different roles played by public authorities and civil promoters – will allow us to focus on several issues that are fundamental to an understanding of ethnic market formation.

Diversity Festival: from political mobilization to ethnoenterprise

Martim Moniz is a large plaza located at the heart of Mouraria, a neighbourhood of 1.65 ha[8] in the historic part of Lisbon's city centre. It dates back to the early medieval period of the city and literally means 'the Moorish quarter'. Literature review shows it has often been, and still is, associated with social exclusion and insecurity (Mapril 2001, 2002a, 2002b), an image still perpetuated by the media.[9] Additional negative stereotypings include illegal goods trafficking, filth, chaos and danger. On top of that, significant parts of the neighbourhood have been demolished since the 1940s because of the poor housing and living conditions. Some shopkeepers were displaced to temporary premises, and the whole area took on an atmosphere of decay and dereliction that persisted until the 1970s.

Such images of danger and degeneration led the Portuguese native population to avoid the area. Significant changes did not occur until immigrants began settling into the neighbourhood in the 1970s. The low rents and public neglect provided favourable conditions for migrant business formation. The first immigrants to start businesses came from Africa, many of whom were of Indian descent (for ethnographic information see Ávila and Alves 1994; Malheiros 1996). As time went by, new groups of traders began replacing the first settlers – the Chinese formed the second large inflow, and Bangladeshis are recent newcomers (see further details on the 'vacancy chain' [Waldinger 1994] in Mapril 2001, 2002a, 2002b).

Specific businesses dealt in all sorts of mass-produced commodities (clothing, eyeglasses, domestic appliances) and local services (hairdressers, restaurants, supermarkets). The former products arrived in large-scale import, mainly from Asian countries, and were then sold further to local consumers or to traders and peddlers such as Gypsies, Pakistanis or Chinese. The service trade provided the burgeoning migrant enclave with many kinds of local support. Martim Moniz has meanwhile become the area with the highest concentration of immigrant entrepreneurs in Lisbon, whose dynamic entrepreneurial activity extends all the way to Spain (Mapril 2001). The press has echoed the evolution of the neighbourhood by conveying images of diverse flavours, colours and fragrances – and good business opportunities – alongside the persistent diffuse image of unsafety.[10]

A similar image of ethnic precincts has been invoked in several initiatives by Lisbon city government in an attempt to combat negative stereotyping. This was part of a wider strategy for Lisbon called the Multicultural and Cosmopolitan Policy, launched in 1993 with the creation of the Conselho Municipal das Comunidades Imigrantes e das Minorias Étnicas (municipal council for the affairs of immigrant communities and ethnic minorities). In 1998, the Lisbon local authority organized the first multicultural carnival in the plaza, featuring Brazilian, Chinese,

Cape Verdean and Mozambican parades (Mapril 2001). This was later discontinued.

In June 2003, another event, the fifth Diversity Festival, took place in Martim Moniz, and the square was again transformed into an urban attraction. Tourists, neighbourhood residents and other city inhabitants made for an unusual gathering. The event was organized by a left-wing anti-racist organization (Rede Anti-Racista, or RAR) and the Portuguese Social Forum, and received support from City Hall, which also was to include it in listings of Lisbon summer festivals.[11] Organizers presumably drew inspiration from activities in other European cities, especially Barcelona. They saw the festival not just as a voice for immigrant communities, but as a lever to open up space for public debate on human rights, anti-racism and immigration. Thirty small stands were erected to sell products ranging from food and drink to 'ethnic' handicrafts. An offer by City Hall to pay for a band performance was turned down, because all other participants were performing for free. Whereas four other such festivals had already taken place at other locations, this was the largest in scale and the first to directly involve City Hall as well as political, migrants' and other voluntary organizations (though only a few migrant businesspeople).

The relatively large number of visitors (put at around 4000 over the four days of the event) was evidence that other, albeit limited, sections of society had been alerted to the event through publicity, in addition to smaller circles of people reachable by word of mouth, such as political activists or migrants seeking contact with co-ethnics. Mediating processes indeed included an advertising campaign partly organized by RAR, with posters produced for the first time in municipal facilities. Non-migrants active in the festival largely came from the Portuguese elite, such as students or artists attracted to 'alternative' lifestyles. People like these have the potential to produce considerable leverage effects. But beyond such visitors and a handful of regular outside customers, the local businesses in Mouraria still form a feeble magnet for consumers looking for new urban experiences framed in ethnic diversity. By and large, the media convey an image of an industrious but inwardly directed enclave.[12]

The public authorities have shown ambivalence. While supporting the festival and some local regeneration initiatives, they balked in the year 2000 when a Chinese entrepreneur proposed creating a Chinatown to attract tourists to Mouraria. The authorities reportedly claimed this would clash with the historic character of this city-centre district.[13]

Although the image of Martim Moniz is slowly changing, it is still hardly seen as a tourist attraction by Portuguese natives. Although two Portuguese guides on 'African Lisbon', produced by an Angolan author in 1993 and 1999 to coincide with the 1994 European Capital of Culture and the Festival of the Oceans events, do mention Mouraria as a point of interest, they are long out of print. At present, only foreign tourist guides highlight it as a precinct of 'ethnic Lisbon' and a visitors' must (Carvalho 2005).

Thus, in terms of an 'ideology of ethnic diversity', it would seem that the Lisbon authorities have meanwhile developed some awareness of the regenerative potential that some of the city's ethnic concentrations may have for the urban fabric, and that, in keeping with a wider trend, the authorities are now acknowledging the marketability of the image of the 'aestheticized Other'. Nevertheless, building a market clearly requires more than just economic demand and immigrant initiatives. Despite the potential benefit of branding Mouraria as an ethnic precinct, its transformation into a real tourist product still seems still far away.

From shantytown to showcase: the Sabura project at Cova da Moura[14]

Sabura is a Cape Verdean Creole word meaning something enjoyable.[15] At Cova da Moura, a shantytown on the outskirts of Lisbon (16.8 ha), the initiative that bears the name Sabura is intended as an experience of 'ethnic tourism', bringing together a series of local ethnic businesses. With attractions like music, restaurants, dance and hairdressing, it outwardly projects an image of Africa.[16] Organizing an initiative like this from a diffuse supply of cultural material helps to open up local immigrant goods and services to the mainstream market of tourism. It thus serves broad economic, sociocultural and political goals. (A partial opening to extralocal demand had already been made, as the 'ethnic supply' gradually spilt over the confines of local demand, especially at weekends.)[17] Promotional materials for the Sabura project state four objectives: to promote a positive image of the *bairro*[18] and its residents in public opinion; to create opportunities to discover the potentials and added value of the bairro; to promote the (African) culture of Cova da Moura; and to develop and stimulate economic activities in the district.

The Sabura project is of very recent origin. It arose out of activities fostered by the local organization Moinho da Juventude ('mill of youth'). The general aim is to help alter the stereotyped image of Cova da Moura as a site of crime, violence and other social ills. Implementation was not without problems, as careful organization was required. Conscious of the need to involve the local community, the new 'tourism agents' and mentors of the project took a survey of local entrepreneurs designed to win them over to the project. They also used it to inventory the ethnic products that qualified for the label of 'Africanness' that Sabura wanted to promote. Reactions to the survey were widely varied, reflecting both support and disapproval. Although the project is still in its fledgling stages at this writing, it has already had a significant impact, particularly in terms of the changing attitudes of some local entrepreneurs who have opened up their spaces to the 'tourist gaze' (Urry 1990) – to the scrutiny of people that had previously been seen as 'strangers to the bairro', and hence were not welcome.

Publicity leaflets show that African cuisine and other services such as 'the art of African hairdressing' and African music and dancing form the bulk of the ethnic supply. Pre-established walking routes are set out in a local guide. The project is also marketed as an opportunity to build bridges – 'a way to show the good things' of the neighbourhood to the outside world. Actually, this process of building a structured supply of African references also aims to build bridges *between* co-ethnics by emphasizing the positive aspects of the immigrant cultures, and thereby fostering a sense of respect and pride for their own 'cultural heritage'. The project has yet another ambition too: to secure the right to permanently remain in a coveted peripheral urban space now under threat from current and projected urban renewal schemes.

The development of this ethnic supply is anchored mainly in the actions of local leaders of Moinho da Juventude, who drew inspiration from a newspaper article about a similar, ongoing experience in the Netherlands.[19] Aware of a growing demand for African cultural manifestations (mainly from co-ethnics, from Portuguese and foreign students, and from people in cultural walks of life), these leaders took the initiative to 'open up the bairro' by way of tourism supply. Before the tourism project began in 2004, the organization had also issued invitations to 'outsiders' on several occasions in an attempt to ease the urban, social and cultural closure of the bairro.

One particular occurrence sent visitors flocking to Cova da Moura: the appearance of a young black singer with a characteristic hairstyle on a popular television show. Media diffusion like this helped build public awareness of Sabura as a space offering a 'little piece of Africa in Cova da Moura'. The official website of the ACIME supported the process by giving publicity to this new tourism product, and later also to the Kola San Jon festival[20] (now included in Sabura), as a form of cultural supply specific to Cova da Moura.

Although most of these products are actually expressions of Cape Verdean culture, the Sabura tourism project uses the 'African' label as an all-encompassing reference. And indeed, though the local majority is of Cape Verdean origin, there are other nationalities living in Cova da Moura too. The bairro serves as a stepping-stone for various groups of new immigrants, who continue to settle there while other, upwardly mobile residents tend to move away, selling or renting their places in Cova da Moura to newcomers. The former residents do keep visiting the neighbourhood. As suggested elsewhere, the bairro thus becomes transformed into a source of inspiration and a point for renewing original ethnic references (see Basu and Werbner 2001).

As a pioneer actor in Sabura, Moinho da Juventude plays a central role in this ethnicization process by helping create social representations of the differences that are manifested in the organized cultural expressions of immigrants. Tourism is a particularly instrumental factor in this process, helping to 'essentialize' cultural references that relate both to African culture in general and to the local culture of the bairro. The supply of 'Africa in Lisbon' operates dialectically, anchored in references both of closure and of openness which sustain the ethnic construction. Towards the outside world, it is a mechanism for image-building; to the inside world, it builds references that can shape subsequent processes of social- ization. The cognitive process of constructing both outward stereotypes and inward mental interpretation schemas derives raw material from this process in which the creation of cultural references is mediated by the tourism project and by the specific interethnic relations it involves. In this process of constructing ethnicity, the roles of the media, of public institu- tions, as well as of local entrepreneurs and organizations are vital, but not without conflict. The Sabura project is still in an embryonic state. In fact, it might better be called an ethnic display, rather than an established ethnic supply. For the process to attain the latter status, new ingredients will need to be added to the 'tourism resource' – notably human resources, tourist infrastructures and tourist intermediaries, such as local tourism develop- ers or wider tourism agencies capable of promoting the new product in the market (Lima da Costa 2004).

In terms of tourism infrastructure, the local supply is still very limited. Demand for accommodation, for instance, has to be referred to hotels in surrounding areas. Another key issue is security. This is a problem still unsolved in Cova da Moura, and it is one reason why organized tours are conducted with a local guide specially trained for the task – he knows which places to avoid, and which persons are willing to act as 'tourees' (van den Berghe 1994) and which are not.[21] Another limitation concerns the poor hygiene in the district, which has led the organizers to pressure the local authority for a more hands-on approach to urban intervention.

To establish intermediaries, contacts have been made with some inter- national agencies, particularly in Germany, Belgium and the Netherlands.[22] The Sabura initiative has already received wide publicity in the Portuguese media, reinforced by personal networking by local organization leaders who have contacts in non-mainstream lifestyles such as artistic and student circles. These represent the chief human resources now involved in operating the project, although more specialized collaboration has also been arranged, notably with a student in tourism studies and with a local resident who works as the local guide.

In a general sense, the Sabura ethnic tourism project is also employed as a means of putting pressure on local government officials – not least in opposition to a citywide urban regeneration scheme which 'amounts to practically demolishing the whole bairro, [a plan] obviously not accepted by local inhabitants'.[23] So far the city government has been unresponsive to such demands. It does not yet acknowledge the potential of the Sabura project for regenerating this urban enclave.

Building a market of ethnic references: some conclusions

It takes more than just supply and demand to build a market. As Alfred Chandler (1978) showed, institutional arrangements have to be created for any market to thrive. The success of the Lisbon Expo '98, a top-down initiative and a major event intended to celebrate diversity, owed much to the massive financial and political investments made by the Portuguese public authorities. Sabura, by contrast, is grassroots initiative, a project owing its existence to a local organization that has succeeded in mobilizing local-level enthusiasm, and one of whose major aims is to put pressure on local government. It has also made some transnational contacts in the tourism industry. The Diversity Festival is likewise rooted in civil society, but it is led primarily by alternative native groups that are active in the political system. They have mobilized local organizations, but they generally keep their distance from the limited support offered by public authorities.

How these three initiatives operate cannot be understood without analysing institutional conditions; the intuitive explanation of the 'invisible hand' does not suffice (see Kloosterman and Rath [2001] on 'mixed embeddedness'). The two latter case studies have additionally shown that the active mobilization of contacts, energies and other resources, largely through informal social networks, is crucial to achieving aims (cf. Guyaz 1981).

As witnessed in the mobilization of civil and political activism that underlies the initiatives, diversity can be celebrated in many different ways. Although ethnoentrepreneurs can be viewed as innovators in the emerging 'symbolic economy' of cities (Zukin 1999), their agenda extends far beyond economic opportunities. Their activities provoke debate about some of the major options facing society. They thereby try to tip the negotiation balance in favour of migrants' cultural and political integration. Both immigrant and native actors may be players on this side of the game. Special mention is due to immigrants' offspring, who resist being channelled into inferior positions and who play a key role in appropriating cultural marginalization as a strategic tool for negotiating a different status.

In the Cova da Moura study, a local organization played a major part in shaping an ethnic we-consciousness, profiting from the bottom-up approach encouraged by EU-funded programmes.[24] In this case, the organization may justifiably be labelled an ethnopolitical entrepreneur (in the sense of Brubaker 2002). Also deserving that label are the local immigrant businesspeople who have consented to move out of the 'we-ghetto' mindset in order to put their neighbourhood to use as an ethnic resource, for both economic and political purposes. Both these groups constitute the basic structure for ethnoentrepreneurship in Cova da Moura.

In the Diversity Festival study, the anti-racist activism fuelling the initiative again demonstrates how ethnic references, used as a banner in the fight against social exclusion and economic exploitation, can be deeply anchored in cultural and political demands. Such demands have a special appeal to certain 'culturally selective' segments of the majority – including many people who are influential in cultural production and innovation (for a similar conclusion in relation to 'Black Paris', see Jules-Rosette 1994). Although the immediate economic benefits of the festival are not significant, some longer-term positive effects for minority integration seem likely.

In both of our case studies, the innovative role of ethnoentrepreneurs lies in their ability to firmly embed their economic aims in wider social and political concerns – and vice-versa. Drawing on their past experiences as activists for alternative social causes or as peripheral or marginalized 'Others', the ethnoentrepreneurs of Lisbon, just as the French service-sector innovators studied by Guyaz (1981), have already overturned some deeply seated beliefs both in economic practice and with respect to the social order sustained by past cultural and political hierarchies.

Other crucial mediating agents include a variety of 'imagineers' (Zukin 1999), who have shown themselves capable of building and conveying an idea of ethnic singularity. Closure and the defining of clear-cut limits between 'us' and 'them' are central to this process. But how can such closure articulate with the building of bridges, which are a condition for the market to exist? Both case studies strongly indicate that servicing the demand for diversity requires an ability to 'invent traditions' (Hobsbawm and Ranger 1983) in ways that are compatible with perceived and aestheti-cized cultural differences. This indeed appears to be one of the conditions for creating demand and sustaining the market.

In Lisbon, the staging of cultural diversity, or simply 'Africa', spear-heads a wider movement. In the case of the Diversity Festival, the slogan is the fight against the unequal status of centre and periphery; in Cova da Moura, the goal is to demonstrate how diversity can enrich the post-colonial setting. In both cases, the construction of cultural diversity hinges on symbols and images of 'us' and 'them', and of both 'openness' and 'closure'. Both contrasts are ideologically meaningful and aesthetically congenial to specific segments of the visiting public.

Innovation in this case can come from a multitude of agents, ranging from politically engaged entrepreneurs to isolated creators of culture. Innovators do not necessarily coincide with ethnoentrepreneurs. The latter, however, have an interest in playing the 'Otherness' card if they want to consolidate the market. And that, in turn, will favour the opening of new economic opportunities in cultural production.

In both case studies, demand was plainly a factor, but watching one's surroundings and what others are doing is also crucial to the functioning of ethnoentrepreneurs (Brewer 1978). This means that their personal networks are crucial assets. Following H. White's (1992) typology, we could say that these markets operate in an arena-type discipline, where innovations may appear in different guises depending on the situation.[25]

How does this kind of supply – framed in particularistic and inwardly directed terms – get matched to the fragmented and diversified demand? As seen in the sociology of art, gatekeepers are pivotal elements; they are active players in the institutional context. Their role in translating the existing supply into an economic asset is decisive. This means re-framing cultural expressions widely seen as 'inferior' into intriguing manifestations of Otherness – that is, into positively valued expressions of difference. Gatekeepers are therefore not just professionals performing a mediating role of bringing together supply and demand; they involve a much wider spectrum of occupations and functions. Following Howard Becker's (1982) insights, one might say that every participant in these 'cultural worlds' is a gatekeeper. Here again, the young people, notably the more educated ones, have major responsibilities in this mediating func tion. Transnational agents may also be crucial players. In both of our case studies, keeping an eye on the market arena also meant attentively monitoring what is being done elsewhere to create and shape demand.

Such ethnoentrepreneurs also play a vital role in shifting from a cognitive framework focused on immigration issues to one focused on the celebration of diversity. Both case studies suggest that the peripheral position of ethnoentrepreneurs in mainstream power relations enables them to mobilize influential relays both within and outside the ethnic groups to act as gatekeepers. The cognitive shift impacts on social life in several ways. At the economic level, it facilitates the development of new economic and occupational opportunities related to the creation of 'ethnic products'. At the cultural level, the valorizing dynamic of the production of cultural diversity may have significant effects in both the social and the political spheres.

Research on the use of ethnic references as a resource to achieve economic aims is now underway as a new and relevant topic in the research agenda on migration (Anderson 1990; Rath 2005). Both the roots-searching drive and the cosmopolitan drive find in the metropolitan context a conducive milieu in which to thrive. Both are important conditions for building a market of ethnic diversity (Halter 2000). The urban context is likely to be the privileged locus for studying such developments. Their effects should be multi-stranded, and should pervade both the cultural and the political dimensions of city life.

'Renegotiating identities' (Kastoryano 2002) through cultural work that more sharply demarcates ethnic categories also carries the risk of tipping the balance in favour of segmentation and closure. A caveat is therefore in order. The creation of 'exotics' may indeed form a condition for promot ing economic incorporation, but it does not automatically result in the systematic social and cultural inclusion of excluded 'Others'. Policymakers should therefore be alert to the emergence of backlash against cultural pluralism (as described by Hollinger 1995), which can easily produce new forms of social exclusion.

Notes

1   This research was funded by the Fundação para a Ciência e a Tecnologia under grant POCTI/SOC/47152/2002. The Luso-American Foundation provided additional support. A note of appreciation is owed to Catarina Oliveira for collecting data for an initial version of this paper, which was presented to the 8th Metropolis Conference in Vienna in 2003, and to the English language editor, who did excellent work. We are particularly grateful to Jan Rath for his insightful comments and suggestions on the first version of the text.

2   'The lusophone community' refers to the worldwide group of Portuguese-speaking countries; on its literary production, see Maciel 2004.

3   We use this term after Brubaker's (2002) 'ethnopolitical entrepreneurs', but with a broader meaning: ethnoentrepreneurs are all the people involved in ethnocultural activities who use or transform these activities into economic and political resources (see also Malik 2002).

4   Ghettoization involves simultaneous exclusion and identity assertion (cf. Maalouf 2001).

5   The Portuguese overseas expansion was the implicit theme of the exposition.

6   This is a term used to denote the post-colonial situation.

7   In January 2003, the Portuguese minister of economy cited ethnic tourism as one of the emergent tourism sectors that merits special attention (Portal do Governo 2003).

8   We thank Marc Latapie for the area measurements.

9   See for example *Suplemento Capital*, 16 July 1983; *Jornal da Região: Lisboa Norte*, 2 October 2000; *Público*, 16 March 2003.

10 R. Felner, 'Martim Moniz aprumou-se para receber Jorge Sampaio', *Público*, 16 March 2003; 'Centro Cultural', *Diário de Notícias*, 10 May 2003.

11 This and the following are data collected in interviews during 2004.

12 *Diário de Notícias*, 16 October 1995; *Independente*, 5 June 1998; *Visão*, 16 March 2000; *Diário de Notícias*, 9 October 2001; *Diário de Notícias*, 10 May 2003.

13 Interview with a member of the Chinese community in 2003.

14 This section is a part of an ongoing PhD thesis by Francisco Lima da Costa.

15 Information gathered in interviews with local informants.

16 The title of the Sabura leaflet is 'Africa, so near'.

17 Up to 3000 visitors per weekend have been estimated by Dias *et al.* (2002); the local population is about 6000.

18 *Bairro*, broadly meaning neighbourhood or quarter, also conveys an idea of a closed community; cf. the Spanish *barrio*.

19 *Diário de Notícias*, 26 August 2003.

20 This is a traditional Cape Verdean festival already held for several years in Cova da Moura.

21 In a newspaper interview (*Público*, 6 March 2004), the local guide compared his experiences to those of guides in South Africa, where ethnic tourism reportedly includes visits to deprived areas where marginality prevails.

22 Interviews conducted with Moinho da Juventude leaders during 2004; see also Silva (2004).

23 Interview with a project participant.
24 The URBAN and EQUAL programmes were mentioned in the interviews. On the structuring effects of opportunity structures, and in particular on the EU-led institutional framework, see Arnaud and Pinson (2004), Rogers and Tillie (2001) and Penninx *et al.* (2004).
25 As Dupuy and Thoenig (1986) showed, selling domestic appliances in Japan, the USA and France involves different modes of market operation, and hence different strategies for innovation.

References

ACIME (Alto Comissariado para a Imigração e Minorias Étnicas) (2005) 'Sabores do mundo: guia da diversidade gastronómica', available http://www.acime.gov.pt/sabores-domundo/ (accessed 25 January 2005).

Almeida, M.V. (2000) *Um Mar da Cor da Terra: raça, cultura e politica da identidade*, Oeiras: Celta.

Anderson, K.J. (1990) 'Chinatown re-oriented: a critical analysis of recent redevelopment schemes in a Melbourne and Sydney enclave', *Australian Geographical Studies*, 28: 131–54.

Anderson, B. (1991) *Imagined Communities: reflections on the origin and spread of nationalism*, London and New York: Verso.

Arnaud, L. and Pinson, G. (2004) 'Shaping the identity and mobilising the "ethnic capital" in three European cities', in F. Eckardt and D. Hassenpflug (eds) *Urbanism and Globalization*, Frankfurt am Main and New York: Peter Lang.

Ávila, P. and Alves, M. (1994) 'Da Índia a Portugal: trajectórias sociais e estratégias colectivas dos comerciantes indianos', *Sociologia: Problemas e Práticas*, 13: 115–33.

Basu, D. and Werbner, P. (2001) 'Bootstrap capitalism and the culture industries: a critique of invidious comparisons in the study of ethnic entrepreneurship', *Ethnic and Racial Studies*, 23(2): 236–62.

Becker, H.S. (1982) *Art Worlds*, Berkeley: University of California Press.

Boxer, C.R. (1963) *Race Relations in the Portuguese Colonial Empire 1415–1825*, Oxford: Clarendon Press; trans. E. Munerato (1977) *Relações Raciais no Império Colonial Português*, Rio de Janeiro: Tempo Brasileiro.

Brewer, J.D. (1978) 'Tourism, business, and ethnic categories in a Mexican town', *Studies in Third World Societies*, 5: 83–102.

Brubaker, R. (1992) *Citizenship and Nationhood in France and Germany*, Cambridge MA and London: Harvard University Press.

Brubaker, R. (2002) 'Ethnicity without groups', *European Journal of Sociology*, XLIII(2): 163–89.

Brubaker, R. (2004) 'Ethnicity as cognition', *Theory and Society*, 33(1): 31–64.

Cabral, J.P. (2001) 'Galvão among the cannibals: the emotional constitution of colonial power', *Identities*, 8(4): 483–515.

Carvalho, F. (2005) 'Os negros na imagem da cidade de Lisboa', SociNova Working Papers, Lisbon: SociNova, Faculdade de Ciências Sociais e Humanas, Universidade Nova de Lisboa.

Castelo, C. (1999) *O 'Modo Português de Estar no Mundo'*, Porto: Afrontamento.

Chandler, A. (1978) *The Visible Hand: the managerial revolution in American business*, Cambridge MA: Harvard Belknap Press.

Cidra, R. (2002) 'Ser real: o rap na construção de identidade na Área Metropolitana de Lisboa', *Etnologia*, 12(4): 189–222.

Collins, J. and Castillo, A. (1998) *Cosmopolitan Sydney: explore the world in one city*, Sydney: Pluto Press.

Contador, A. (2001) *Cultura Juvenil Negra em Portugal*, Oeiras: Celta.

Crane, D. (ed.) (1994) *The Sociology of Culture*, Oxford: Blackwell.

de Seabra, H. (2005) *Delinquência a Preto e Branco*, Lisbon: Fim de Século.

Di Leonardo, M. (1998) *Exotics at Home: anthropologies, others, and American modernity*, Chicago and London: University of Chicago Press.

Dias, J. *et al.* (2002) *Concurso 'Housing for the poor': Cova da Moura/UN*, mimeo, Faculdade de Arquitectura, Universidade Técnica de Lisboa.

DiMaggio, P. (1997) 'Culture and cognition', *Annual Review of Sociology*, 23: 263–87.

Dupuy, F. and Thoenig, J.C. (1986) *La Loi du Marché*, Paris: L'Harmattan.

Ferreira, C. (2002) 'Processos culturais e políticos de formatação de um mega-evento', in C. Fortuna and A.S. Silva (eds) *Projecto e Circunstância*, Porto: Afrontamento.

Fischer, C. (1999) 'Uncommon values, diversity and conflict in city life', in N. Smelser and J. Alexander (eds) *Diversity and Its Discontents*, Princeton NJ: Princeton University Press.

Fortuna, C. and Peixoto, P. (2002) 'A recriação e reprodução de representações no processo de transformação das paisagens urbanas portuguesas', in C. Fortuna and A.S. Silva (eds) *Projecto e a Circunstância*, Porto: Afrontamento.

Fradique, T. (2003) *Fixar o Movimento: representações da música rap em Portugal*, Lisbon: Dom Quixote.

Guarnizo, L.E. and Smith, M.P. (1998) 'The locations of transnationalism', in M.P. Smith and L.E. Guarnizo (eds) *Transnationalism from Below: comparative urban and community research, volume 6*, New Brunswick NJ: Transaction Publishers.

Guyaz, J. (1981) 'Innovations dans le secteur des services: du militant à l'entrepreneur', in J.D. Reynaud and Y. Grafmeyer (eds) *Français, Qui Êtes-vous?* Paris: La Documentation Française.

H2T (2003) 'Hip-hop tuga', available http://www.h2tuga.net (accessed 27 April 2003).

Halter, M. (2000) *Shopping for Identity: the marketing of ethnicity*, New York: Schocken.

Hargreaves, A. (ed.) (2004) *Minorités Postcoloniales Anglophones e Francophones*, Paris: L'Harmattan.

Hargreaves, A. and McKinney, M. (eds) (1997) *Post-Colonial Cultures in France*, London and New York: Routledge.

Hobsbawm, E. and Ranger, T. (1983) *The Invention of Tradition*, Cambridge: Cambridge University Press.

Hollinger, D. (1995) *Post-Ethnic America*, New York: Basic Books.

Ireland, P. (1994) *The Policy Challenge of Ethnic Diversity*, Cambridge MA: Harvard University Press.

Jules-Rosette, B. (1994) 'Black Paris: touristic simulations', *Annals of Tourism Research*, 21(4): 679–700.

Kastoryano, R. (2002) *Negotiating Identities: states and immigrants in France and Germany*, trans. B. Harshav, Princeton: Princeton University Press.

Kloosterman, R. and Rath, J. (2001) 'Immigrant entrepreneurs in advanced economies: mixed embeddedness further explored', *Journal of Ethnic and Migration Studies*, 27(2): 189–202.

Laplantine, F. and Nouss, A. (2001) *Métissages*, Paris: Pauvert.

Lima da Costa, F. (2004) 'Turismo étnico, cidades e identidades: espaços multiculturais na Cidade de Lisboa: uma viragem cognitiva na apreciação da diferença', SociNova Working Papers, available HTTP: <www.fcsh.unl.pt/socinova/migration> (accessed 19 November 2004).

Maalouf, A. (2001) *In the Name of Identity: violence and the need to belong*, trans. B. Bray, New York: Arcade.

Maciel, C. (2004) 'Língua Portuguesa: diversidades literárias: o caso das literaturas africanas', SociNova Working Papers, available http:www.fcsh.unl.pt/socinova/ migration (accessed 19 November 2004).

Malheiros, J. (1996) *Imigrantes na Região de Lisboa*, Lisbon: Colibri.

Malik, B. (2002) 'Arrival of the native: ethno-entrepreneurship gets cracking as the well-settled generation of Non-Resident Indians begins to accumulate culture', available http://www.himalmag.com/2002/september/review_2.htm (accessed 6 January 2005).

Mapril, J. (2001) 'Os Chineses no Martim Moniz: oportunidades e redes sociais', SociNova Working Papers, available HTTP: <www.fcsh.unl.pt/socinova/migration> (accessed 20 September 2004).

Mapril, J. (2002a) 'De Wenzhou ao Martim Moniz', *Ethnologia*, 12–14: 253–94.

Mapril, J. (2002b) 'Transnational jade formations or the translocal practices of Chinese immigrants in a Lisbon innercity neighbourhood', in F. Eckardt and D. Hassenpflug (eds) *Consumption and the Post-Industrial City*, Frankfurt am Main and New York: Peter Lang.

Mariza (2004), available http://www.mariza.org (accessed 20 April 2004).

Marques, M.M., Mapril, J. and Dias, N. (2003) 'Migrants' associations and their elites. building a new field of interest representation', SociNova Working Papers, available HTTP: <www.fcsh.unl.pt/socinova/migration> (accessed 20 September 2004).

Marques, M.M., Dias, N. and Mapril, J. (2005) 'Le "retour des caravelles" et la lusophonie: de l'exclusion des immigrés à l'inclusion des lusophones?' in E. Ritaine (ed.) *Politiques de l'Etranger: l'Europe du Sud face à l'immigration*, Paris: Presses Universitaires de France, pp. 149–83.

Penninx, R., Kraal, K., Martiniello, M. and Vertovec, S. (eds) (2004) *Citizenship in European Cities: immigrants, local politics and integration policies*, Aldershot: Ashgate.

Portal do Governo (2003), available http:www.portugal.gov.pt/pt/Conselho+de+Ministros/ Documentos/20030122MEcBTL.htm (accessed 10 November 2003).

Público (2004a) '"Mariza", uma personagem incontornável', available http://www.publico.pt/ sites/fado2004/002Mariza1.asp (accessed 20 April 2004*)*.

Público (2004b) 'Rui Vieira Nery: na linha do fado', available http://www.publico.pt/ sites/fado2004/01RuiVieiraNery.asp (accessed 20 April 2004).

Rath, J. (2005) 'Feeding the festive city: immigrant entrepreneurs and tourist industry', in E. Guild and J. van Selm (eds) *International Migration and Security: opportunities and challenges*, London and New York: Routledge.

Rogers, A. and Tillie, J. (eds) (2001) *Multicultural Policies and Modes of Citizenship in European Cities*, Aldershot: Ashgate.

Rosa, M.J.V., de Seabra, H. and Santos, T. (2004) *Contributos dos 'Imigrantes' na Demografia Portuguesa: o papal das populações de nacionalidade estrangeira*, Lisbon: Observatório da Imigração.

Santos, B.S. (2001) 'Entre Prospero e Caliban', in M.I. Ramalho and A. Ribeiro (eds) *Entre Ser e Estar*, Porto: Afrontamento.

Sassen, S. and Roost, F. (1999) 'The city: strategic site for the global entertainment industry', in D.R. Judd and S.S. Fainstein (eds) *The Tourist City*, New Haven and London: Yale University Press.

Silva, M.M. (2004) *Jornal de Notícias*, http://ladoesquerdo.blogs.sapo.pt/arquivo/ 2004_03.html (accessed 26 March 2004).

Taylor, C. (1994) 'The politics of recognition', in A. Gutmann (ed.) *Multiculturalism*, Princeton NJ: Princeton University Press.

van den Berghe, P. (1994) *The Quest for the Other: ethnic tourism in San Cristóbal, Mexico*, Seattle: University of Washington Press.

Vermeulen, H. (1999) 'Immigration, integration and the politics of culture', *Netherlands Journal of Social Science*, 35: 6–22; trans. (2001) *Imigração, Integração e a Dimensão Política da Cultura*, Lisbon: Colibri.

Urry, J. (1990) *The Tourist Gaze: leisure and travel in contemporary societies*, London: Sage.

Waldinger, R. (1994) 'The making of an immigrant niche', *International Migration Review*, 27(1): 3–30.

White, H. (1992) *Identity and Control*, Princeton NJ: Princeton University Press.

Yeary, M.J. (2003) 'American musicians, French music: the formation of the jazz culture in post-war France', available http://home.uchicago.edu/~yeary/papers/amfm.pdf (accessed 29 January 2005).

Zukin, S. (1999) 'Whose culture? Whose city?' in R.T. LeGates and F. Stout (eds), *The City Reader*, New York: Routledge.

11 Tourists 'R' Us

Immigrants, ethnic tourism and the marketing of metropolitan Boston

Marilyn Halter

On a sunlit Friday afternoon in late September, Michele Topor guides a dozen eager visitors through the bustling streets of Boston's Little Italy. Topor's inventive North End Market Tour is one of several excursions available in the metropolis that promoters have come to dub America's Walking City – Boston, Massachusetts. Famous for its restaurants and cafés, pastry shops and salumerias, the North End delights all the senses. With her insider's knowledge of this historic Italian American community, Topor feeds the intellect as she imparts her culinary expertise and gastronomic advice on selecting aperitifs, digestifs and grappas. The excursionists follow Topor, single file, along the North End's narrow sidewalks. They pass a sunny corner where a group of older, carefully dressed men, all with hats on and some with canes, have set up their folding chairs. The men are talking, playing cards and revelling in each other's company. In the course of a day, the tour guide comments, they will move from one vantage spot in the neighbourhood to the next, following the warmth of the autumn sun.

The tour group is comprised primarily of young to middle-aged couples. Two coincidentally are from Houston, but have never met before; two other couples (each with a stroller and toddler in tow) do know each other but are from opposite coasts – from Sacramento, California, and from Framingham, Massachusetts, a town less than 30 miles away – while the rest of the group is also drawn from the Greater Boston area. Four of the sightseers have Italian roots, while the tourists with the young children are Filipino, speaking Tagalog to each other and English with everyone else. The Filipinos found out about the jaunt by watching a special on the Food Network, the others through a North End website or rave reviews from neighbours. Topor does not advertise at hotels and other such outlets, but instead relies solely on word-of-mouth and the media, which have been very generous to her. Features on television, radio and in various publications both nationally and internationally have drawn the 3000 people who sign up annually. Indeed, last year she had to hire an assistant to help handle all of the interest. At $42 (€35) a person and lasting three-and-a-half hours, the tour attracts a committed and adventurous clientele.

These are visitors to the city who want an alternative to the standard tour package, seeking greater authenticity, meaning and connection in their travels.

As urban tourists looking for an authentic experience of Italian ethnicity, they have found the perfect venue by joining the North End Market outing. Or have they? It turns out that Topor has no Italian ancestry, but is actually a Polish American, whom some locals still view with suspicion even though she has lived in the North End for more than 30 years. Nor do most proprietors of the many Italian-themed dining spots, groceries and gift shops reside in the North End; they live in the outskirts of the city. Paradoxically, the gentrification that triggered the mass exodus of Italian Americans from the North End in the 1970s coincided with the development of the neighbourhood into a destination saturated with Italian ambience. At its height in the 1930s, 95 per cent of North End residents claimed Italian heritage, many of them first-generation; today's figure is less than 40 per cent. And the elderly men on the corner? At the end of the day, with one exception, the men will load up their chairs into the trunks of their Toyota Camry or Ford Taurus sedans and drive back to the suburbs of Medford or Everett, Woburn or Revere where they actually live. Yes, they did grow up in the North End, maybe even started a family there, but in the 1970s and 1980s, as real estate values and their economic circumstances changed, they moved. Some wanted their own back yards and barbecues; others, at a later stage of life, were nudged by their grown children to leave the old neighbourhood and relocate closer to the kids in a suburban home. Retired now and perhaps widowed, they have become commuters again – fighting off the isolation and anomie of suburbia. These days they navigate the notorious Boston traffic not to get to their jobs on time, but for the simple pleasures of conversation and companionship on the streets of the North End. On display here, these ex-North Enders are, and are not, the real thing. But the very same can be said for the sightseers. According to industry standards, only those travelling from a distance of 50 miles (80 km) or more are actually classified as tourists, a criterion that disqualifies the touring locals who make up half the group.

Addressing the issues raised by this trek through the North End, this chapter explores the changing contours of ethnic tourism and the marketing of immigrant cultures in an urban setting – Boston, Massachusetts, New England's largest city. What has been the relationship of immigrants and ethnics to the tourism industry during the last thirty years? How have the dynamics of the commodification of traditions evolved, and what has the impact been on the economic culture of the metropolis? What is authentic and what commercialized? Furthermore, who are the tourists and who are the toured?

When the Ellis Island[1] website debuted two years ago, digitising the museum's immigration records, the foundation that runs it was taken completely by surprise by the magnitude of traffic seeking information – it averaged more than 25,000 hits per second. Similarly, the Boston Public Library reports that over 50,000 people visit it every year sheerly to trace their genealogies – further evidence that not only is the United States a nation of nations, but contemporary Americans are a nation of roots-seekers. A full-blown, multi-faceted ethnic revival has been in motion in the United States for more than three decades now, propelling American-born descendants of immigrants to actively re-identify with their ethnic heritages. Thought initially to be a passing fad, the so-called 'roots' phenomenon now seems stronger than ever. It is manifested in the growth of ethnic celebrations, a zeal for genealogy, and a greater interest in ethnic artefacts, cuisine, music, literature and language. Even the renowned baby and child care expert, Dr T. Barry Brazelton felt moved to proclaim: 'Every baby should get to know their heritage'. In the fundamentals of twenty-first-century American childrearing, roots training comes even before potty training (Halter 2000).

A central element in the contemporary preoccupation with tracing and celebrating roots has been the escalation of interest in ethnic tourism – travelling to sites either to explore one's own ethnic history and identity or to learn more about the cultural backgrounds of other people. A study of US adults who travelled in the past year found that an extraordinary 81 per cent, or 118 million, fell under the classification of historic/cultural travellers. The report also showed that on average this group spent more money per visit than other sightseers, further evidence of the lucrative possibilities in the business of ethnic tourism (TIA 2003). As Sharon Zukin (1995: 2) has emphasized in her research on urban landscapes, 'culture is more and more the business of cities,' and as such ethnicity is playing a central role in the revitalization of urban centres. Thus, as more and more travellers and vacationers seek out ethnocultural experiences, they find themselves drawn to centres of immigrant commerce, whether in Europe, Australia or North America. Because of the differences in the histories of immigrant flows, however, such activity in the United States – home to long-standing settler populations – tends to revolve around ethnic-, rather than specifically immigrant-generated, tourism; while by comparison in Europe, where concentrations of foreign-born newcomers tend to be more recent, the likelihood of direct immigrant involvement in the tourism industry is much greater. Typically, in municipalities across the United States, a mix of immigrant-, second-, third- or even fourth-generation residents are coalescing to showcase the cultural diversity of their ethnic neighbourhoods, whether Little Tokyo or Little Italy, in a concerted effort to capture tourist dollars. While visitors seek a connection to the city that will enable them to feel like residents and blend with the locals, immigrant and ethnic entrepreneurs use those tourist dollars for urban socioeconomic development.

Exotics at home

Tourism as a search for authenticity used to be associated with travel to faraway lands and resort vacation locales to experience the so-called exotic. In their quest to experience difference, first-world tourists have sought encounters with the 'authentic' in third-world sites (van den Berghe 1994; Desmond 1999; MacCannell 1976, 1992; Selwyn 1996; Wood 1998). John Urry's (1990: 11–12) groundbreaking study, *The Tourist Gaze,* named these dynamics, arguing that

> potential objects of the tourist gaze must be different in some way or other. They must be out of the ordinary. People must experience particularly distinct pleasures which involve different senses or are on a different scale from those typically encountered in everyday life.

Taking the appeal of difference as the basis of ethnic tourism, I argue here that the kaleidoscopic shifts in the ethno-racial make-up of many regions of the United States create a new set of circumstances whereby the 'exotic' today can be found much closer to home, in the immigrant and ethnic neighbourhoods of metropolitan areas, while the dynamics of ethnic tourism shift significantly within the urban market. From Santa Fe, New Mexico, to Frankenmuth, Michigan, cultural exploration that used to be associated with foreign travel is now on the agenda of domestic tourists keen to experience the out-of-the-ordinary closer to their own backyards. Meanwhile, such trends are galvanising local planning agencies, which see the new ethnic tourism as an opportunity to spark economic development and which recognize the importance of selling authenticity in the process (Satterthwaite 2001). Using Boston as a case study, this chapter interrogates distinctions between tourists and the toured as well as the meaning of ethnic tourism itself, showing how these categories and concepts have become more complex and blurred in post-industrial settings.

Heritage tourism meets ethnic Boston

Greater Boston is an ideal site for exploring these dynamics. The metropolis is a central player in the global service economy, receives significant flows of immigrants and attracts tourists in large numbers. During the past two decades, this traditional Yankee city, one of the oldest and most densely populated settlements in the United States, has been profoundly transformed from a largely white community of people of European descent to a variegated multicultural metropolis, and from 'an old-fashioned mill-based economy to a highly diversified mind-based one'.[2] Indeed, Boston contains what is likely the world's largest concentration of academic and research institutions. Many foreign newcomers to the region today are just as likely to arrive by BMW as by boat; these professionals work in the universities, laboratories and research hospitals that form the basis of the city's high-tech future. Other newcomers, carrying the requisite tattered bundles over their shoulders, are desperate to join the low-skilled service economy.

Totalling close to 600,000 residents in the city proper (and 3.5 million in the Greater Boston metropolitan area), Boston became a majority minority city in the 1990s for the first time, as Latinos, Asians and immigrant blacks arrived and as tens of thousands of whites departed. Currently, one out of four residents was born in another country, while 140 different languages are spoken in the public schools. Yet, unlike most US cities with large concentrations of the foreign-born, in Greater Boston no single ethnic group predominates. And while Boston's centres of high-tech glitz and its networks of global communication at first appear to be worlds apart from the grime and grit of the inner-city neighbourhoods (oftentimes located just a few blocks away), there are nevertheless some striking similarities. Both populations are cosmopolitan in their ethnic make-up, and both bustle with enterprise and vitality. Thus, highly skilled professionals, entry-level workers and holiday travellers all gravitate to this multi-layered metropolis, where international business, knowledge and consumption networks are intricately linked. The image of the staid, homogeneous city that Oscar Wilde once dubbed 'the paradise of prigs' is being refashioned, almost overnight, as a dynamic urban centre that dazzles with diversity.

Prior to 1980, Boston indeed had a reputation for dealing very poorly with diversity. As a consequence of racial conflicts and turf wars, and in particular the school busing crisis of the 1970s, many people had come to associate the metropolis with intolerance and bigotry, rather than with the values of liberty and equality so often linked to the patriotic consciousness imbued in the city's Colonial heritage. After all, Boston was the first city in America to abolish slavery; yet by the 1970s it was the intransigence of the city's racial divide that was making national headlines. By contrast today, as part of the embrace of diversity, Boston is one of just two urban centres in the country (the other, not surprisingly, is New York) that has a special mayor's office dedicated to its newest residents. Created in 1998, the Office of New Bostonians was formed to give immigrants access to government resources and to get them invested in the civic and cultural life of the metropolis. At the fourth annual New Bostonians Community Day, billed as the city's most visible celebration of immigrants and multi-culturalism, a City Hall banner welcomed attendees in six languages. A recipient of the Outstanding New Bostonian Award, Nigerian-born Kate Chinwe Okoye, closed her acceptance speech with the advice, 'Your culture gives you your identity. Share it – pass it on to your children'.[3] Another recent City Hall initiative was the publication of a new visitors' pamphlet, *Beyond baked beans: discover Boston main streets* (City of

Boston 2004), featuring 19 diverse neighbourhoods to explore. Whether the visitors taste *frijoles negros* in Egleston Square or sip on a Vietnamese bean milkshake at Fields Corner, the mayor entreats them to experience 'the Real Beantown' not covered in traditional tourist guidebooks.[4]

As the most 'old-world' of US cities and steeped in its Colonial past, Greater Boston has long been marketed as the ultimate destination for American heritage tourism. While ethnic and heritage tourism are usually treated as very different aspects of the larger industry, Boston's ethnic tourism interfaces and overlaps with its already well-developed Colonial tourism sector. Tourism represents the third largest industry in the Commonwealth of Massachusetts, raking in $9 billion (€7 billion) in direct spending and tax revenues in the year 2003. It also employed 147,000 people (MOTT 2003). A survey based on the nearly 13 million domestic and international tourists to Greater Boston in the year 2000 ranked visiting historical places and attending cultural events and festivals second and third respectively (behind shopping) as the most frequently reported trip activities for travellers. New publications such as *Boston's neighborhoods: a food lover's walking, eating and shopping guide to ethnic enclaves in and around Boston* (Morgenroth 2001) attest to the growing appeal of ethnic tourism and its promise of invigorating the local economy. For example, although it took more than 200 years of Irish presence in the city for even one tourist guide focused on this population to appear (the *Guide to the Boston Irish* [Quinlin 1985]), the current visitor to the greater Boston area can find hundreds of print and online resources pointing to Irish cultural events. Perhaps the most concerted recent effort to promote ethnic tourism of any kind in the Boston area was the formation of the Boston Irish Tourism Association (BITA) in the year 2000. With Massachusetts claiming the highest percentage per capita in any state (23 per cent) of residents professing Irish ancestry, the agency has already distributed half a million brochures featuring local Irish businesses and cultural events.

More fledgling groups look to the established ethnic populations as models for their own efforts. In their promotional literature, the organizers of a new Center for Latino Arts, in conjunction with the already existent Jorge Hernández Cultural Center, express their hope to

> become a fixture engraved into the cultural life of Boston as a whole. Like Chinatown and the 'Little Italy' of the North End, a unified Latino Cultural District will become a destination in tourist guides. It will be both a place that Latinos can call their own and a place where non-Latinos will feel welcomed and talk about when they return home. Moreover, it is believed that the Center will become an integral part in developing an 'Arts Village' in Boston's South End, which will foster economic development through increased tourism.
>
> (IBA 2003)

Whether contributing to ethnic cookbooks, tracing their family trees or creating new ethnic organizations, individuals are intent on finding ways to personally express their ethnic identities. Perhaps the most common public display of this romance with ethnicity has been the increasing popularity of the ethnic festival. At the nexus of commerce and culture, such activities are typically the outcome of a combination of commercial, civic and cultural resources; in recent times, they have become paradigmatic events to celebrate cultural diversity. The short film on Boston's Ethnic Festivals that ran continuously at the city's Dreams of Freedom immigration museum encouraged visitors to 'learn about Old World Culture and enjoy New World shopping and cuisine'.[5] Indeed, from Boston's St Patrick's Day extravaganza and Greek Independence Day festivities in the springtime to Portuguese feasts and Caribbean carnivals throughout the month of August, along with Italian Saints' Day celebrations all summer long, tourists have myriad opportunities to participate in such events. In the market world of the ethnic festival, it is not at all unusual for corporate, small-business, penny-capitalist and non-profit interests to be brought together for the enjoyment of locals and tourists alike. They are an opportunity for co-ethnics to celebrate their own cultural heritage, to raise public awareness to their group and to open a window to outsiders into another culture. At their best, they function to simultaneously raise money and multiethnic consciousness.

Ethnic tourism is not a completely new phenomenon. As early as the beginning of the twentieth century, it had become popular for the white upper and middle classes to go 'slumming', and to shop in New York's or San Francisco's Chinatown or in Chicago's 'Ghetto Market'. Guidebooks, graphically illustrated with sketches of the unfamiliar terrain, were published to help the sightseers navigate unknown streets (Cocks 2001). In the early twentieth century, many visitors to Boston identified the immigrant North End – reputedly second only to Calcutta in population per square metre – with the worst of American slums. Nevertheless, if they wanted to pay tribute to one of the heroes of the American Revolution, they could not avoid its seamy streets. The early 1900s had spawned a new movement to preserve the Colonial heritage of the city, and first on the list was an initiative to save Paul Revere's house, located in once-elegant North Square, then the very centre of the immigrant ghetto.

> Generations of antiquarians and tourists made the pilgrimage through a district they found alien and threatening, where they endured, said one, 'the vile odors of garlic and onions' to gaze at the greatly altered structure sporting a commemorative plaque and a 'variety of signs with Italian names', which the *Boston Globe* found, 'strangely out of harmony with the memories of its earlier history'.
>
> (Holleran 1998: 216–17)

Whereas, earlier in the twentieth century, the basis for attempts to refashion North American cities to make ethnic enclaves more appealing to tourists was to look to the ideal of the European city and to recast immigrants as non-threatening old-world peasants, today the situation is reversed. European municipalities see the metropolitan areas of the United States, Canada and Australia – with their long-standing immigrant communities and the resources of culturally diverse populations – as models for their own initiatives to capitalize on the commercial potential of newly burgeoning ethnic neighbourhoods.

Nowadays, poor immigrant precincts formerly meant to be avoided are instead objects of civic pride. During the course of that transition, a reversal of sorts occurred when a 1970 edition of the *Boston Passport* guide warned a minority group, African-Americans, to stay out of the North End at night; by then the reputation of the neighbourhood had shifted from alien slum to a bastion of white, working-class patriotism:

> If you are black ... you would be well-advised to venture farther into the North End only by daylight. Remember that the North End, besides being the home of Sacco and Vanzetti[6] also harbored the poet who penned [the patriotic song] *My country 'tis of thee*: to be black or long-haired is now considered un-American. North Enders are friendly and helpful – but intolerant and tough.
>
> (Lockwood 1970: 22)

The turning point in the city's recognition of its multiethnic constituencies ironically coincided with the elaborate preparations for the US Bicentennial events of the mid-1970s. Alongside Revolutionary War and Yankee Colonial heritage attractions, guidebooks began to underscore the immigrant legacies of the metropolis. Boston's *Official bicentennial guidebook* included a lengthy essay entitled 'Changing Boston' (Lupo and Rivers 1975), documenting the municipality's ethnic past, block-by-block and neighbourhood-by-neighbourhood, shifting the focus from monuments and museums to the contributions of people who actually worked and dwelled in the city. In addition, beginning in the spring of 1975, the Mayor's Office of Cultural Affairs developed an ethnic calendar of special events, including not only the expected Irish, Italian, Jewish and 'Afro-American'[7] months but a month each dedicated to Boston's Armenian, Albanian, Lithuanian, Hispanic, Chinese, Haitian, Greek, Arab, Portuguese, Polish and Native American populations. The series concluded with Multicultural America Month, December 1976, where, in an unlikely reinvention, the featured activity was Boston Tea Party Day.[8] Thus the rhetoric of cultural pluralism began to permeate the tourist literature, even when the sites themselves were fundamentally tied to the city's Colonial history. Most of the guidebooks published before the 1960s refer to 'Historic Boston' in their titles. By the 1970s, not only is the 'historic' dropped, but the content has also changed to include information about ethnic festivals such as Chinese New Year, August Moon and Italian Saints' Day Feasts as well as other attractions located in immigrant neighbourhoods.[9]

By 1930, a 'must' stop for Boston tourists, the Union Oyster House (in operation since 1826 and considered the city's oldest restaurant) was advertised in the Grey Line bus tour guide, *Souvenir of Boston* (1930). Interestingly, the caption under a photo of the establishment reads 'Ye Olde Oyster House' – this was the height of New England's Colonial Revival, after all, and the guide touts the historical significance of the site, letting prospective customers know (among other titbits) that the very stalls and oyster bar remain in their original positions; that the building itself dates to 1717, when one Thomas Capen sold imported, fancy dress goods there; and that Louis Philippe, later to become king of France, lived on the second floor of the building during his exile, teaching French to leading Bostonians – a very early example of multiculturalism seeping into Brahmin Boston. Today a much more prominent display of diversity is the location of the six graceful glass towers, etched with 6 million numbers that make up the Holocaust Memorial, dedicated in 1995. In the heart of historic Boston, the monument stands directly across from the Union Oyster House and adjacent to Faneuil Hall, a structure given to the town of Boston in 1742 by another Frenchman, Peter Faneuil. Today it forms the benchmark (pioneered by James Rouse, the legendary Boston developer) for the adaptive re-use of crumbling historic market buildings to create tourist-oriented commercial plazas. Rouse's concept of festival marketplaces helped to catapult 'ye olde' Boston into a pulsating post-industrial city.[10]

Yet another attraction a stone's throw from Faneuil Hall, the Union Oyster House and the Holocaust Memorial is Curley Park, where twin lifelike statues comprise a monument commemorating the four-time mayor of Boston, as well as state governor and congressman, James Michael Curley, the son of Irish immigrants from Galway who dominated local politics during the first half of the twentieth century. A short walk to historic Boston Common brings the visitor to the nineteenth-century Robert Gould Shaw Memorial, a tribute that further exemplifies multiethnic representation in a major city landmark. Whether following the Black Heritage, Irish Heritage or Freedom Trail, tourists will arrive at this monument to the Black soldiers of the Civil War's storied 54th Regiment led by Colonel Shaw. Augustus Saint Gaudens, who immigrated to Boston in 1848, fleeing the Irish famine with his French father and Irish mother, is the sculptor responsible for this commemoration.

The placement of significant landmarks associated with ethnic minorities, such as the Holocaust Memorial or the Curley monument, in close proximity to historic sites along Boston's famed Freedom Trail provides graphic visual testimony to how the city's immigrant heritages and its more recent multiethnic consciousness have been incorporated together to reshape the actual built environment and established urban landscapes of the city. Created in 1951 and reputedly the world's first heritage walking tour, the Freedom Trail serves as a grand tribute to Boston's Yankee past as it links and preserves 16 key markers related to the birth not only of the Colonial revolution, but of the nation itself. Yet today visitors can wend their way through these classic sites of US patriotism from several different paths and perspectives – following the maps of the Jewish Friendship, Black Heritage or Irish Heritage walking tours, all of which intersect and overlap significantly with stops along the Freedom Trail in today's multicultural Boston. Just as vernacular life in ethnic neighbourhoods which were still deemed slums in the recent past is now being commodified and reinvented, so too has the marketing of the built environment led to a shift from a monolithically Anglo cultural landscape to a polyglot array of public landmarks.

The immigrant theme park

By the 1980s many urban immigrant enclaves had been transformed into tourist attractions resembling theme parks. For scholars, urban planners and merchants, such shifts open up the question of precisely how such a fragile equilibrium of commerce and culture can be maintained. During the past several years, the North End's Annual Fisherman's Feast, first celebrated in 1911, has had to seek corporate sponsorship to keep the oldest continuous Italian festival in Boston alive. The Sicilian fishing fleet, the festival's traditional source of funding in addition to local businesses, had been in serious decline over the last two decades, compelling feast organizers to accept major financial support from the Sorrento Cheese Corporation. This US company had chosen an overall marketing goal of achieving greater exposure in the Italian-American community, as well as a specific branding strategy to associate their label with the perception of a 'genuine Italian cheese'. The transition from primarily localized community support to multinational corporate sponsorship of the Fisherman's Feast was initially quite smooth. From the company's point of view, the partnership has been an unqualified success. Through various promotions such as a celebrity 'cheese-building contest', which attracted Boston area sports and television personalities of Italian descent, as well as the decision to make key charitable contributions, Sorrento sharpened its image as a community-focused corporate citizen, while also drawing substantial media attention. In return, the feast leadership was relieved to have the resources to continue staging all the elaborate events of this three-day extravaganza. But planning for the 2003 celebration ran into a stumbling block. Sorrento escalated its involvement by cementing its logo right onto this cultural gathering – changing the festival's name, after nearly a century's existence, to the 'Sorrento Cheese Fisherman's Feast'.

For many in the local community, a line had been crossed – from recognition to exploitation. Not surprisingly, Sorrento prevailed, but many North Enders are still feeling bruised by the controversy.

Such levels of commercialization raise important questions about whether the marketing of sites for ethnic tourism is a beneficial trend or whether it leads to what David Harvey (1985: 150), in his examination of urbanization and capitalism, has termed 'creative destruction'. Does commodification of experience render it meaningless? And in what ways does ethnic tourism serve to combat ethnocentrism and promote an appreciation of diversity; or is it, instead, just another vehicle to accentuate cultural stereotypes, and such an inauthentic experience that it becomes yet another example of the disneyfication of American society and culture? Sometimes the marketplace can actually foster greater awareness of ethnic identity, and can even act as an agent of change in that process.

Consider the case of Cajun identity. These descendants of the Acadians, French colonists who settled along the chilly Bay of Fundy in the early seventeenth century, had been exiled by the late eighteenth century to semitropical Louisiana bayou country. After more than two centuries of adaptation to US society, the Louisiana Cajun community had lost many of its distinctive cultural features. In the mid-1970s, however, a steady stream of tourists began to arrive in the area looking for an 'authentic' Cajun experience. They came, in particular, from Francophone regions like Quebec, France and Wallonia in Belgium, inspiring the largely assimilated local population to probe its own cultural uniqueness. Indeed, many of the Francophone tourists took a stronger interest in Cajun culture than the natives themselves did. At the instigation of these tourists, Cajun identity was re-awakened in Louisiana. A fully-fledged cultural revival has since taken place, and visitors can now find restaurants serving traditional and reinvented Cajun fare, ethnic festivals featuring Cajun music, exhibitions that carefully trace the history of migration and settlement, and renewed attention to distinctive language usage. Given the sweeping ethnic renaissance that has occurred nationwide over the last three decades, Louisiana Cajuns might have eventually moved in this direction on their own initiative. Nonetheless, tourism was instrumental in accelerating this renewal of ethnic consciousness (Esman 1984).

Even more dramatic is the role that the tourist industry is now playing in unifying a long-fractured Ireland. In part to reverse the recent downturn in tourist visits resulting from the 11 September attacks in the US and the outbreaks of foot-and-mouth disease in Irish cattle in 2001, the Northern Ireland Tourist Board and the Bord Fáilte, the tourist board of the Irish Republic, recently formed a single agency known as Tourism Ireland to jointly promote the entire island as one tourist destination. Both sides of this formidable north–south divide recognize the potential economic benefits of marketing a unified Ireland to international visitors, especially to North American tourists, who make up a sizable proportion of travellers to the island. The hope is that such significant commercial cooperation, as well as the symbolic implications of presenting Ireland as a single destination, will spill over into the political arena and even spur the peace process. On the North American side of the Atlantic, tourism officials are eager to see Tourism Ireland succeed. Particularly when it comes to Massachusetts, and especially the city of Boston, travel is very much a two-way street. Visitors from Ireland and Britain come to see the Massachusetts sites as well, more than half a million annually, and they represent a constituency that is being actively courted by both the Massachusetts Office of Travel and Tourism and the Greater Boston Convention and Visitors Bureau. Just like any other tourist, these Europeans are seeking cultural, shopping and entertainment experiences. Visits to the JFK Library, the Irish Cultural Centre and well-known Irish pubs in the city such as the Black Rose (established in 1976) are frequent stops in their travels (Quinlin 2002).

In the post–modern metropolis, even at sites that appear at first glance to be indelibly stamped by their particular ethnic heritage, cultural hybridity and multiple identities predominate. At the Feast of St Anthony, the largest of the almost weekly series of summertime festivals in honour of the patron saints of the Italian villages that had once been home to the immigrants, it is actually hard to find an old-fashioned Italian sausage stand among the cultural mosaic of vendors hawking beignets and henna art, teriyaki and piña coladas. Even the woman selling the Canoli Girl and Italian Princess T-shirts is of Asian descent.

Moreover, while there is no question that the event does celebrate Italian traditions among these other representations of global cultures, the organizers are just as eager to demonstrate their local pride in Boston. The iconography of the city appears at many junctures – from the use of the Boston Bruins hockey team flag as the centrepiece of the Italian American donor banners, to the abundance of Red Sox posters and pennants, and even to the conflation of Boston baked beans with Italian coffee beans in a display of classic beanpot souvenirs alongside the signature Lavazza Italian coffee makers. With spectacle now such an integral factor in the marketing of post-industrial cities, a mix of 'high' and 'popular' cultural activities, professional sports events and venues, and the performance of ethnicity (invented and otherwise) creates a tantalising and potentially profitable commodity for tourists to consume.

Reversing the gaze

The notion of alienated leisure, whereby tourists distance themselves from those they tour by asserting their socioeconomic, racial and ethnic status to maintain established hierarchies of power and privilege, was much more pronounced in earlier periods than it is today. In the nineteenth century and well into the twentieth, sightseers in the United States were mostly native-born, Anglo and wealthy, while the gaze was on the most 'acceptable' local population of foreigners or their descendents (Shaffer 2001;Whiteside 1926). But what happens to the long-standing notions of the sociocultural dynamics of ethnic tourism – notions that hold to a fixed paradigm of an assumed white, moneyed tourist, partaking of the cultural representations of ethno-racial minorities as the exoticized toured – when those doing the touring are themselves minorities? As major metropolitan areas in the United States become more and more diverse through large-scale migration, and especially as minority populations gain greater socioeconomic success, such groups are directing their buying power at a range of tourist venues, just as their Anglo counterparts have long done (Halter 2000). The new publication *Multicultural Travel News* recognizes this shift. In addition to helping tourists and business travellers add a multicultural agenda to their visits to various cities across the country – by designing brochures that highlight heritage-specific trails, cultural festivals and historical neighbourhoods – the magazine also explicitly encourages Hispanic, African-American and Asian-American travellers to go and explore these sites. In this instance, the 'ethnic' in ethnic tourism simultaneously characterizes both the types of destinations and the diverse heritages of the travellers themselves.

Today, members of a tour group may just as well be first- or 1.5-generation immigrants as native-born. For example, Don Quijote Sightseeing Tours reverses the gaze by specialising in showing Latino visitors the sites of Yankee New England. The family-operated Boston business, offering services in Spanish and English, takes pride in the dignitaries who have used their agency, including King Juan Carlos I of Spain and movie actor Edward Olmos. What does the tourist gaze fix on, then, during a Don Quijote excursion? Or consider the new glossy travel magazine, *Odyssey Couleur*, aimed at racial minorities, with its clever subtitle *Travel in Color* and its motto 'Real People, Real Travel'. The publication's founder, Linda Spradley Dunn (2003: 8), introduces the premier issue by explaining,

> As a frequent traveler, I spend plenty of time in airports perusing travel, adventure and leisure magazines. But as I thumb through their pages, I'm always disappointed by how few images I see of multicultural tourists. There are photos of 'natives' smiling obligatorily and modern-day versions of bare-breasted women. Yet, I'm often left wondering, 'Where are the people like me?' Even though major publications overlook us, I know that people of color are traversing the world. I see us in every airport I pass through, striding down the concourse, phones pressed to our ears, PDA's ready, laptops slung over our shoulders.

Dunn refers to her target audience as 'globetrotters of color' and claims that these Black and Latino tourists constitute the fastest-growing segment of the travel industry. The Bermuda Department of Tourism must agree. In a recent full-scale promotion to boost leisure travel to the tropical island, it hired a well-known multicultural marketing firm, Images USA, to focus on reaching African-American consumers.

Newcomers all

At another level, R. Timothy Sieber's recent study of the revitalization of downtown Boston found that currently as many as half of the city's families have been living in the metropolis for less than five years. As is the case in many other post-industrial municipalities, many of the residents are newcomers. This is due not only to an escalation in foreign-born arrivals, but also to high mobility rates within the US population itself and to migration back into the city by those who had previously moved to its outskirts. Sieber (1997: 68) observes that 'local residents – especially professionals – and suburban commuters enjoy consuming the same representations of the city's "authentic" history, heritage and local color that are the mainstay of tourist programming'. Thus, whether tourists or residents, they are seeking similar types of nostalgic ambience and authenticity of experience. Indeed, a survey of 656 people was conducted in 2000 to gauge visitor interest in the proposed Boston Museum Project, an attraction that represents the first city-and-regional museum. Half the participants were 'local residents' (residing within the Route 495 beltway) and half were 'tourists' (residing outside it). Local residents indicated a greater interest in topics such as Boston's diverse population, cultures, immigration and neighbourhoods, while the tourists were primarily interested in Boston's role in the American Revolution (Greif *et al.* 2000: 10). City Hall understands this trend, too, as reflected in the mayor's endorsement of the *Main Streets* guide, where he concludes, 'Whether you are coming from across the country or across the street, I invite you to explore our neighborhoods, meet our residents and discover the Boston "Beyond Baked Beans"!' (City of Boston 2004: 2).

In the past, guidebook rhetoric took on a tone of tolerant racism, portraying ethnic Others as part of the allure of the picaresque, alongside the then-prevailing assumption of darker skin and fundamental inferiority. Thus, they were simultaneously appreciated and degraded. Today the ideology of multiculturalism permeates tourism literature, emphasising the greater value of difference and the enriching aspects of diversity. Even though the language of difference is now more acceptably conveyed, the appeal of authenticity remains constant. The tendency to exoticize still lurks. Despite going so far as to specifically acknowledge the ethnocentrism of Boston's past, the author of a recent glossy city guide with all the right multicultural spin, still introduces photos of the small community of Tibetans in Cambridge, Massachusetts, by declaring, 'From its intolerant Puritan origins, greater Boston has come to be the home of exotic international cultures' (Marcus 1998: 10). In such depictions – which are quite commonplace in the urban cultural tourism literature – racial and ethnic minorities are emphatically valued and insidiously marginalized at once.

In the end, of course, by far the largest numbers of immigrants connected to the contemporary tourism industry are the behind-the-scenes labour force that includes armies of marginalized and low-paid service workers who clean hotel guestrooms, wash dishes in restaurants and scrub toilets in museum restrooms. Or they may be construction workers that physically restore the designated historic sites and build the newest attractions on the urban cultural landscape. Their own impact on the contours and direction of urban tourism, however, is negligible, and the tourist gaze passes right through them. At the same time, tourism and hospitality services provide necessary and crucial first jobs for many newcomers, refugees and immigrants alike. This is a sector of the economy that willingly hires people from a wide range of backgrounds and speaking many different languages. Indeed, in the housekeeping department at the Omni Parker House, America's oldest continuously operating hotel located right on the Freedom Trail, the staff has posted a world map covered with tiny flags that represent the many different countries of origin of the employees. Such employment patterns explain why the International Institute of Boston, a refugee resettlement agency also providing legal, social and language services, operates its only systematic direct job training and placement programme in the hotel industry. The Ethiopian refugee, arriving penniless and traumatized after months or years of severe deprivation in the process of finally reaching the United States, may start within six months of arrival as a hotel service worker, after twelve more may find work as a parking garage attendant, and within three to five years may inch up the ladder to the night shift at a convenience store. These are all low-end positions, to be sure, yet it was cleaning up after the tourists at hotels that formed his entrée into the workforce. Such a job may be typically viewed as dead-end, but not by everybody. Many people in refugee and immigrant communities regard the slightest increment of mobility, even within the unskilled labour pool, as a significant achievement.

Alas, tourists long to feel like residents, and residents hunt for local tour guides. Filipinos enrol in Italian culinary expeditions, while Spanish royalty saunter down the Irish Heritage Trail. Meanwhile the locals at a Saints' Day parade display fingers in ruddy henna designs. With such an astounding interpenetration of cultures and economies, of globalized marketing campaigns and multiply identified constituencies, traditional categories and interpretations no longer suffice. We are all doing the gazing and we are all being gazed upon. In the current construction of ethnic tourism, tourists are everywhere and tourists 'R' us.

Notes

1 Ellis Island in New York harbour was the principal receiving station for European immigrants to the United States from 1892 to 1954.
2 B. Bluestone and M. Huff Stevenson, 'Boston in bloom: the region's "triple revolution" has fed a growth spurt', *Boston Globe*, 30 July 2000.
3 Author's field notes, 17 September 2003.
4 Boston baked beans are a widely known traditional Boston specialty.
5 As an exception to the success of other immigrant tourism sites in Boston, the Dreams of Freedom museum had to shut its doors in September 2003 due to lack of funding. The former director, Westy Egmont, suggested that an explanation for its demise could be that those ethnic museums and centres in the city that represent one particular population, such as the recently established Irish Cultural Centre, are usually successful by virtue of established ethnic networks in which a few key underwriters can bring in major support. The Immigration Museum was too diffuse, too broadly construed to rally the necessary dedication and ongoing funding. The only comparable site to succeed is the Ellis Island Museum, which has federal support and represents everyone's immigrant ancestors (conversation with Westy Egmont, 15 October 2003).
6 Sacco and Vanzetti were two Italian immigrant anarchists who were tried and executed in a controversial case in 1921.
7 In the 1970s, the preferred label for the African-American population was 'Afro-American'.
8 The Boston Tea Party was a rebellious event in 1773 that marked the start of the Colonial revolution against the British.
9 Prior to the 1970s, if restaurants with foreign cuisines were listed at all, they were classified as 'international dining', which usually meant Continental, French, Italian, Greek, Polynesian or Chinese restaurants, and which (except for the Chinese places) were not tied to local immigrant populations. From that time on, dining recommendations began to be categorized by ethnicity and were often listed as part of the draw to tour immigrant neighbourhoods.
10 In late 2003, in a move to again revamp Faneuil Hall and bring even greater attention to the city's historic and local attractions, the mayor of Boston announced that some of the first-floor specialty shops would be forced out to make room for a US National Park Service visitors' centre, with exhibits showcasing Boston's neighbourhoods (D. Slack, 'Faneuil Hall makeover would oust some merchants', *Boston Globe*, 2 December 2003).

References

City of Boston (2004) *Beyond Baked Beans: discover Boston main streets*, Boston: City of Boston. Online. Available http://www.cityofboston.gov/mainstreets/PDFs/BMS_Beans2004_2.pdf (accessed 1 February 2005).

Cocks, C. (2001) *Doing the Town: the rise of urban tourism in the United States, 1850–1915*, Berkeley: University of California Press.

Desmond, J. (1999) *Staging Tourism*, Chicago: University of Chicago Press.

Dunn, L.S. (2003) 'Founder's letter', *Odyssey Couleur*, 1(1): 8.

Esman, M. (1984) 'Tourism as ethnic preservation: the Cajuns of Louisiana', *Annals of Tourism Research*, 11(3): 451–67.

Greif, R., Lunedei, J. and DuRei, S. (2000) *Boston Museum and National Park Project Visitor Intercept Survey Results*, Boston: JSI Marketing Services.

Halter, M. (2000) *Shopping for Identity: the marketing of ethnicity*, New York: Schocken.

Harvey, D. (1985) *Consciousness and the Urban Experience: studies in the history and theory of capitalist urbanization*, Baltimore: Johns Hopkins University Press.

Holleran, M. (1998) *Boston's 'Changeful Times': origins of preservation and planning in America*, Baltimore: Johns Hopkins University Press.

IBA (Inquilinos Boricuas en Acción) (2003) 'IBA starts construction of the Casa de la Cultura, center for Latino Arts', available http://www.iba-etc.org/claContent.htm (accessed 1 February 2005).

Lockwood, R.E. (ed.) (1970) *Boston Passport*, Cambridge MA: Owl Publishing.

Lupo, A. and Rivers, C. (1975) 'Changing Boston', in Boston 200 Corporation (ed.) *The Official Bicentennial Guidebook*, New York: E.P. Dutton Co.

MacCannell (1976) *The Tourist, a new theory of the leisure class*, New York: Schocken.

MacCannell (1992) *Empty Meeting Grounds: the tourist papers*, London: Routledge.

Marcus, J. (photos by S.C. Kelly) (1998) *Boston: a citylife pictorial guide*, Stillwater MN: Voyageur Press.

Morgenroth, L. (2001) *Boston's Neighborhoods: a food lover's walking, eating and shopping guide to ethnic enclaves in and around Boston*, Guilford CT: Globe Pequot.

MOTT (Massachusetts Office of Travel and Tourism) (2003) *Massachusetts Travel Industry Report*, available http.//www.massvacation.com/research (accessed 1 March 2005).

Quinlin, M. (1985) *Guide to the Boston Irish*, Boston: Quinlin Campbell Publishers.

Quinlin, M. (2002) 'Uniting around tourism for the sake of Ireland', *Boston Business Journal*, 22(12): 16.

Satterthwaite, A. (2001) *Going Shopping: consumer choices and community consequences*, New Haven: Yale University Press.

Selwyn, T. (ed.) (1996) *The Tourist Image: myths and myth making in tourism*, Chichester: Wiley.

Shaffer, M. (2001) *See America First: tourism and national identity, 1880–1940*, Washington DC: Smithsonian Institution Press.

Sieber, R.T. (1997) 'Urban tourism in revitalizing downtowns: conceptualizing tourism in Boston, Massachusetts', in E. Chambers (ed.) *Tourism and Culture: an applied perspective*, Albany: State University of New York Press.

'Souvenir of Boston' (1930) Grey Lines Bus Guide.

TIA (Travel Industry Association of America) (2003) *The Historic/Cultural Traveler*, available http://www.tia.org/Pubs/pubs.asp?PublicationID=16 (accessed 1 February 2005).

Urry, J. (1990) *The Tourist Gaze: leisure and travel in contemporary societies*, London: Sage.

van den Berghe, P. (1994) *The Quest for the Other: ethnic tourism in San Cristóbal, Mexico*, Seattle: University of Washington Press.

Whiteside, C.W. (1926) *Touring New England on the Trail of the Yankee*, Philadelphia: Penn Publishing.

Wood, R.E. (1998) 'Touristic ethnicity: a brief itinerary', *Ethnic and Racial Studies*, 21(2): 218–41.

Zukin, S. (1995) *The Cultures of Cities*, Cambridge MA and Oxford: Blackwell.

Index